THE HUMAN TUTORIAL DIALOGUE PROJECT:
Issues in the Design of Instructional Systems

COMPUTERS, COGNITION, AND WORK

A series of volumes edited by:

Gary M. Olson, Judith S. Olson, and Bill Curtis

FOX • *The Human Tutorial Dialogue Project*
MORAN and CARROLL • *Design Rationale*
SMITH • *Collective Intelligence in Computer-Based Collaboration*

THE HUMAN TUTORIAL DIALOGUE PROJECT: Issues in the Design of Instructional Systems

BARBARA A. FOX
University of Colorado, Boulder

CRC Press
Taylor & Francis Group
Boca Raton London New York

CRC Press is an imprint of the
Taylor & Francis Group, an **informa** business

First Published by
Lawrence Erlbaum Associates, Inc., Publishers
10 Industrial Avenue
Mahwah, New Jersey 07430

Transferred to Digital Printing 2009 by CRC Press
6000 Broken Sound Parkway, NW Suite 300, Boca Raton, FL 33487
270 Madison Avenue New York, NY 10016
2 Park Square, Milton Park Abingdon, Oxon OX14 4RN, UK

Copyright © 1993 by Lawrence Erlbaum Associates, Inc.
All rights reserved. No part of the book may be reproduced in any form, by photostat, microform, retrieval system, or any other means, without the prior written permission of the publisher.

Library of Congress Cataloging-in-Publication Data

Fox, Barbara A.
   The human tutorial dialogue project / Barbara A. Fox.
     p.   cm.
   Includes bibliographical references and indexes.
   ISBN 0-8058-0826-4 (c)—ISBN 0-8058-1072-2 (p)
     1. Instructional systems—Design. 2. Interaction analysis in education. 3. English language—Discourse analysis. 4. Intelligent tutoring systems. 5. Tutors and tutoring. I. Title.
LB1028.38.F69   1993
371.3′078—dc20   93-8130
                                                                                                         CIP

Publisher's Note
The publisher has gone to great lengths to ensure the quality of this reprint but points out that some imperfections in the original may be apparent.

Contents

Preface	**ix**
1. Introduction	1
2. Methodology	10
3. Tutoring Dialogue Structure	15
4. Openings	31
5. Correction in Tutoring	51
6. Interaction as a Diagnostic Resource in Tutoring	68
7. The Target of Tutoring	82
8. Bandwidth	94
9. Indeterminacy and Rules	107
10. Conclusions	118
References	**120**
Author Index	**125**
Subject Index	**127**

*The royalties from this book
will be donated to the
Nature Conservancy of Hawai'i.*

Preface

The goal of this book is to begin to document the dialogue processes in naturally occurring human–human tutoring, in the context of informing the design of intelligent tutoring systems, and of interactive systems in general.

Exactly how a study of human–human dialogue can or should influence the design of interactive instructional systems, however, is now a controversial issue. On one view, interactive instructional systems should simulate at least certain facets of human–human communication. On this view, because of the need to model human dialogue, the study of human–human instructional dialogue is seen as a critical step in the further development of intelligent tutoring systems (ITS), in that such research provides the theoretical foundation from which working systems can be built. Something close to this perspective can be seen in the following quote from Galdes (1990):

> . . . we should not spend our time looking for results which say, "Intelligent computer systems will never be as good as humans because humans do <X> and computer systems can't do <X>. Instead, we need to think in terms of *what we can mimic in the human's behavior that seems to make the interaction flow more smoothly*. (p. 373)

Galdes made a crucial point in this statement: Even though human–computer dialogue may differ on the surface in extreme ways from human–human dialogue, at least some (if not most) of the basic underlying *processes* of human–human dialogue must be mimicked (i.e., simulated) in a human–computer interface if it is to be user friendly. Thus, even if users communicate with computers in a way that suggests they believe the computers to be less than adequate communicative partners, some of the processes of natural human–human dialogue will still be apparent. In spite of the potential differences, then, between human–human dialogue and human–computer dialogue, on this view the study of human–human interaction will provide necessary insight into designing simulations within human–computer interfaces (see also McTear, 1987).

There is another view, however, that suggests that interactive instructional systems cannot, at this stage in our development of systems, be simulations of human–human communication; rather, on this view, computers are best seen as dynamic tools for human users, tools that do not attempt to approach the functionality of human tutors. From this perspective, studies of human communication remain useful in serving as guides in developing more user-friendly tools. Hayes (1983) articulated this position well:

> In brief, we believe that direct simulation of human conversation will not play an important role in user-friendly interfaces until speech processing has made sufficient technical advances to allow spoken language to be used freely and in conjunction with pointing, but that the study of human communication will continue to be relevant to interactive computer interfaces of all kinds. (p. 230)

A similar note is sounded by Suchman (1987):

> Some researchers in human–computer interaction make the claim that cognitive science and computer technologies have advanced to the point where it is now feasible to build instructional computer systems that are as effective as experienced human tutors . . . In contrast to this optimism, I have argued that there is a profound and persisting asymmetry in interaction between people and machines. . . . Because of the asymmetry of user and machine, interface design is less a project of simulating human communication than of engineering alternatives to interaction's situated properties. (p. 185)

The project reported on in this study, the Human Tutorial Dialogue Project, is neutral on this general issue. This project is rooted in the belief, shared by many in the artificial intelligence (AI) community, and reflected in both the quote from Galdes and the quote from Hayes, that the study of human communication will provide important insights for the design of interactive systems, whether those systems intend to be simulations or not. The findings of this study thus have a potentially wide range of application; the implications for system design drawn from the analyses presented here will differ according to the theoretical and practical concerns of specific readers and their research programs. The findings reported here are offered as a theoretical and empirical foundation on which such design decisions can be based.

The goal of the project, which forms the basis of this book, was to document and analyze various interactional mechanisms of (more of less) naturally occurring human tutorial dialogues. Originally the project was concerned with the following questions:

- How does the domain of the tutoring (e.g., chemistry vs. physics) affect the dialogue?
- How do changes in bandwidth affect the dialogue?
- How does the nature of the tutor (whether human or "computer") affect the dialogue?
- What was the structure of these tutoring dialogues like?
- How was tutor intervention handled when students were having trouble?
- How was communicative repair handled in the dialogues?

To answer these questions, we[1] videotaped face-to-face, terminal-to-terminal, and simulated "ITS" tutoring sessions, in four science and mathematics domains. A few of these original questions are not addressed in the present study; the interested reader should consult Fox (1990) for those. Moreover, many questions were added to this list as the project progressed; these newer questions are for the most part tackled here.

This project represents the first empirical study of human tutorial dialogue from a conversation-analysis perspective, that is, where the conversational interaction is the focus of analysis rather than, for example, larger scale techniques for teaching. The motivation for this focus comes from two sources: First, although all tutoring systems have implicit theory or theories of minute-level interaction built into them, little research has been done to form an empirical foundation for such theories, and hence current systems tend to be based on the designers' intuitions rather than on data. This fact almost certainly makes systems unnecessarily brittle in actual use. Second, of the small but growing collection of empirical studies of tutoring, almost all have been designed and carried out by computer scientists, whose training naturally leads them to be concerned with interaction at the level of knowledge transfer and teaching techniques. My training as a linguist, on the other hand, brings my attention to the minute-by-minute details of the interaction, in particular to the processes that bring the interaction into existence and allow it to develop relatively smoothly.

It is also an outgrowth of my training as a linguist and conversation analyst that the tutoring dialogues recorded for the current study were created out of the pressing real-life needs of each student. The students who participated in the study were genuinely seeking tutoring in one of the four domains, and the

---

[1]In describing the life and management of the project, I use first person plural pronouns to refer to the group of people, who, in addition to myself, worked on various aspects of the project over the course of the last 5 years. In describing the writing up of this and past reports, I use first person singular pronouns because the writing process has been a solo effort.

student–tutor pairs were given no instructions about what topics to discuss or how to discuss them. So, although in certain respects the setting was not entirely natural—a small lab room with a video camera—we have come very close to tutoring as it occurs naturally at the university level. This fact proves to be useful to the design of interactive systems in that we are thereby given a window on tutoring as it occurs spontaneously, with minimal direction from the researcher. We can then see processes that might otherwise be hidden from view, as in the case of how global plans for the tutoring session are managed, how next problems are selected, how correction is negotiated, and so forth.

But this book is not primarily for linguists or other scholars working in the area of conversation analysis; rather, it is primarily for researchers in the areas of intelligent tutoring systems and human–computer interaction whose job it is to design and build interactive systems; designers have typically had little research on which to base their design decisions and have perforce designed systems with unexplored assumptions regarding the nature of such interaction informing the design. The central goal of this book, then, is to give designers of intelligent interactive systems, in particular of intelligent tutoring systems, both an explicit theoretical framework and empirical findings on human–human tutorial dialogues that can serve as guides for future system designs.

Although the book is oriented primarily toward designers and theorists in the areas of human–computer interaction and intelligent tutoring systems, the focus on the collaborative construction of knowledge should also prove of interest to scholars in education, and to those concerned with social learning of all sorts, including facets of child development.

ACKNOWLEDGMENTS

The Human Tutorial Dialogue Project began in the summer of 1985, with the help of a Junior Faculty Achievement Award from the University of Colorado, Boulder. In the 7 years that have passed since that time, many people have given time and energy to the project, and to the many reports that have come out of it, including this book.

The idea for this work originally grew out of discussions with Gerhard Fischer and his group at the University of Colorado. His early support of this project was essential in getting the research funded and underway.

After the pilot study in 1985, the project was supported by the Office of Naval Research, grant number N00014-86-K-0105. I would like to thank Susan Chipman for recognizing the value of a project like this one and for being perpetually supportive of this project. Thanks also to the many participants of the yearly ONR contractors' meetings who provided helpful suggestions as the project developed.

The principle research assistant on the project was Manuel Arce-Arenales, now professor of computer science at the University of Costa Rica. Without his help and support this project would never have come to fruition. Lorraine Karen and Liang Tao also made valuable contributions to the project at various stages. Rebecca Burns-Hoffman and Jule Gomez de Garcia contributed time and ideas during the pilot project.

The Linguistics Department and the Institute of Cognitive Science at the University of Colorado provided important resources, including accounting expertise, lab space, and professional help with the computer set-up. Many thanks especially to Alan Bell, Gladys Bloedow, Gary Bradshaw, Walter Kintsch, Taki Meghjee, Martha Polson, John Roberts, and Sue Stortz for their help. Special thanks to Susanna Cumming for being such a wonderful colleague and friend.

Getting these ideas into published form has taken effort and patience from several people. Hollis Heimbouch, the editor at Lawrence Erlbaum Associates when the manuscript was originally accepted, was extremely helpful and provided the crucial first round of comments that brought this manuscript from a collection of technical reports into book form. The new editor, Amy Pierce, has continued the tradition. Lucy Suchman has been unendingly encouraging of this work, including providing invaluable comments on a previous version. An anonymous reviewer also made useful suggestions. The final version was written while I was on sabbatical; thanks to the University of Colorado, especially to Dean Middleton, for making that possible.

We were extremely fortunate in having Emanuel Schegloff as a consultant on the project. His genius for interactional analysis never ceases to amaze me, and his insights can be found throughout this work. He was particularly important in tuning the analyses in chapters 5 and 6.

Thanks go to all the friends who helped to keep me going during the life of this project, especially to Anne Hockmeyer, for her understanding and her encouragement of these and other efforts, and to Sandy Thompson, for all of her professional and emotional support. And to my family, my parents in particular, goes much appreciation for their continued support and understanding.

And finally, to the tutors and students who volunteered to be part of this research, for their willingness to be open and to teach and learn in front of our video cameras—many humble thanks.

Barbara A. Fox

1 Introduction

THEMES OF THE STUDY

Although this study reports a variety of specific findings on human–human tutoring, there are three global findings that serve as the major themes of the book.

The first and most crucial of these findings concerns the function of tutoring. The basic hypothesis of this study, grounded in the analyses presented in chapters 4 through 8, is that the function of tutoring (in science and mathematics)[1] centers on the collaborative contextualizing of abstract symbols and descriptions— including linguistic phrases in problem statements, formulae, constants, variables, and so on—to enable the student to project these abstract forms into a series of activities in concrete objects. By contextualization I mean the situating (in Suchman's, 1987, terms) of symbolic forms into a specific, local context, for a particular occasion (e.g., preparing for an exam), for a particular student, within particular constraints (e.g., only an hour for the session), within a particular discourse history. *Collaborative* here refers to the fact that this contextualization is achieved by effort from both the tutor and the student, with initially more active effort from the tutor, but with the ultimate goal of enabling the student to navigate the practices learned in tutoring in a solo journey (e.g., taking a test). The claim here is not only that tutoring itself is situated activity, but also that one of the main goals of tutoring is to teach the student how to situate otherwise

---

[1] Whenever I refer to *tutoring* in this study, I mean to restrict my focus to tutoring in science and mathematics. Research on tutoring in the humanities remains to be done, and our results do not necessarily hold for those domains.

abstract and a-contextual forms. Tutoring thus represents an ideal research arena within which to observe not only how people behave in situated activity but how they teach and learn about situating.

A brief discussion may be useful here in beginning to clarify the notion of contextualization. Consider, as a starting point, the following passage:

(1)
T: So let's do one of these. (1.3) We'll start with the simple ones. Five ex minus twelve over ex times ex minus four. First check the degrees> the numerator is smaller than the denominator so that's okay> factor (0.2) they already did that // for us.

S: Right.

T: And then when you break it up (0.3) this stuff (1.1) is going to have to be equal t- for *ex* we have *one* power, so ey over ex

T: We want these two things to be equal (0.7) so- the first thing we could do is just make them look more alike (0.7) get the common denominator. (0.3) This is a step that you will skip after a few times, but this is (0.2) where it comes from.

In this passage, the tutor takes an abstract form [the equation $5x - 12/x*(x - 4)$] and projects it into a series of actions that she and the student can jointly perform on a collection of concrete objects: The objects are representations drawn on a piece of paper; the actions are first glossed as general procedures (e.g., "check the degrees"), abstract statements that are themselves in need of situating, and then the tutor indicates how to apply these procedures in the given context (e.g., "the numerator is smaller than the denominator so that's okay"). It is this process, roughly described here, that I refer to as the contextualization of abstract forms; and it is this process that I claim is the *raison d'etre* of tutoring.

The second major finding of the study is the indeterminacy of the language of tutoring dialogues. That is, in tutoring, a given sentence does not uniquely specify a single interpretation, but is in principle open to an indefinite number of interpretations and reinterpretations.

Let us look briefly at an example: In one of the physics sessions, the student is trying to set the forces of two electrons equal to the weight of either at the earth's surface. In describing how she had tried to work this problem earlier (before coming to the session), she said:

(2)
S: So I tried to look at the *weight*,

And the tutor responds,

(3)
T: And all's they give you is the m*a*ss

As an obvious case of indeterminacy, the student's use of the term *weight* is vague and without precise interpretation, and the sense of the term changes over time as she comes to see that mass and weight are different, and that weight is mass times g, and therefore that weight is a force.

A less obvious case of indeterminacy, not "fixed" by teaching the student the "correct" meaning of a term, lies in the phrase "tried to look at." In many contexts, trying to look at something implies that one was unsuccessful in actually seeing that thing: One can imagine trying to look at a cell under a microscope, or trying to look at a planet through a telescope. But we (and the tutor) must infer that the problem here is not a visual one, but some kind of metaphoric mental inspection. This inference, and the inference that the inspection fails (i.e., that the student does not in fact "find" the weight), are, of course, contextually driven and are not given in the surface linguistic items of the clause. Furthermore, exactly what this other kind of inspection is is entirely unspecified, left as it is to the tutor and student to work out, if they should find that necessary, or left vague if a more specific sense is unimportant.[2]

It follows from this fact about the language of tutoring dialogue that there is no single *intended* interpretation for a sentence, because the precise sense of an utterance in a dialogue is open to further negotiation as the dialogue proceeds, a process that can reconstruct what appeared to be the intent of the speaker at the original moment of utterance (see Fox, 1987a, for a discussion of reconstruction). The student's use of the term *weight* in the previous example illustrates this phenomenon. The principle of indeterminacy, especially its theoretical status in the study of everyday conversation (see Heritage, 1984), and the related notion of rule-governed behavior, are addressed at length in chapter 9.

The third global finding of this book is that tutoring involves constant, and local, management. This requires a pervasive mutual orientation between tutor and student, such that every session (indeed, every utterance) is a thoroughly interactional achievement, produced by both tutor and student.

THEORETICAL BACKGROUND

One of the recent introductory textbooks for the growing field of cognitive science describes communicative interaction in the following way:

---

[2]But an exact and fixed sense can never be achieved. See chapter 4 for a discussion of this fact.

> Think about your last conversation with a friend. From a certain point of view it went this way. Your friend had an idea or thought that she wanted to convey to you. For this purpose she sent volleys of commands to scores of muscles in her abdomen, chest, throat, tongue, and lips. The resulting oral gymnastics had the effect of vibrating the air molecules around her. Ultimately the vibrations in the air caused corresponding vibrations at your eardrum. These vibrations were passed along in attenuated form through bones, fluids, and tissues in your ear until they resulted in a volley of sensory discharges along the auditory nerve to your brain. The sensory discharges then acted in such a way as to cause some counterpart of your friend's idea to be formed in your brain. This idea gave rise to an idea of your own, so you sent volleys of commands to scores of muscles. . . . Speech and understanding are effortless and unconscious activities that go on while we think about the topic of the conversation. . . . (Osherson & Lasnik, 1990, p. xii)

Although this description is clearly not intended to represent a complete scientific theory of communication, it exhibits assumptions common in the scientific community about communicative interaction that bear on the discussions in subsequent chapters, and hence are worth some investigation.

It appears in this description that communication is a tossing back and forth of ideas, each of which originates in the brain of a single speaker, and which recreates itself in the brain of the other speaker. This process of exchange is effortless and automatic, and it seems that the words themselves (or, rather, the vibrations of air particles they produce) are responsible for "carrying" ideas between speakers. There seems to be no need in this view for mechanisms of interpretation.

This view of communication is widespread, not just in cognitive science but in western society in general: Harris (1981) referred to it as the "telementation" model of communication; Reddy (1979) described it as the "conduit metaphor" (see also Duranti, 1984; Lakoff & Johnson, 1980). Reddy found evidence in common expressions that English offers its speakers a view of communication in which thoughts or ideas are transferred from one brain/mind to another via the conduit of words, and Harris viewed this model used by linguists in their analyses.

It is not surprising to find that speakers of English tend to theorize about communication in a manner that reflects this basic metaphor. This tendency is true for "just plain folks" (Lave, 1988) as well as for linguists, philosophers, and members of the cognitive science community.

It is natural, indeed unavoidable, then, that the theory and design of interactive systems is rooted in the conduit metaphor. Suchman (1987) described communication in one such system:

> To compensate for the machine's limited access to the user's actions, the design relies upon a partial enforcement of the order of user actions within the procedural sequence. This strategy works fairly well, insofar as a particular effect produced by

the user (such as closing a cover on the copier) can be taken to imply that a certain condition obtains (a document has been placed in the machine for copying) which, in turn, implies a machine response (the initiation of the printing process). In this sense, the order of user and machine "turns," and what is to be accomplished in each, is predetermined. The system's "recognition" of turn-transition places is essentially reactive; that is, there is a determinate relationship between certain uninterpreted actions by the user, read as changes to the state of the machine, and the machine's transition to a next display. By establishing a determinate relationship between detectable user actions and machine responses, the design unilaterally administers control over the interaction, but in a way that is conditional on the actions of the user. (p. 108)

In this system, an action by the user creates a corresponding significance in the system—a correspondence determined in advance by the designer—and the system responds accordingly. The user, at least from the system's perspective, then engages in an identical, predetermined, process. There is no interpretation on the part of the system. As Suchman pointed out, this design works well in many instances but produces trouble in others. Suchman traced these troubles, at least in part, to the predetermined, nonsituated, nature of the system's "understanding" of user actions and its responses; and she proposed an interpretive, situated perspective on action and meaning as a more appropriate model.

Suchman's work shares a variety of assumptions and methods with the present study. Like Suchman, the present study departs from the conduit metaphor of communication in assuming that meaning is collaboratively constructed by all of the participants in an interaction rather than conceived in an individual mind and subsequently transferred to the mind of another. In this view, a next speaker's response to a preceding utterance provides not only a display of that next speaker's understanding of the utterance; it also serves as a resource for the first speaker to elaborate on his or her own understanding of the utterance. Further, the significance of turns in an interaction (be they verbal or nonverbal) is not determinate; it is interpreted, by means of a vast array of common sense methods and practices, for the purposes of a particular context, on a particular occasion, and is contingent. The significance can change retroactively (Fox, 1987), be elaborated, or repaired subsequently.

Like Suchman's work, this study finds its theoretical and analytic roots in two paradigms: *ethnomethodology,* especially Conversation Analysis (see, e.g., Atkinson & Heritage, 1984; Garfinkel, 1967; Heritage, 1984; Pomerantz, 1975; Sacks, Schegloff, & Jefferson, 1974; Schenkein, 1978; Suchman, 1987); and *activity theory* (e.g., Giddens, 1979; Lave, 1988; Leontiev, 1981; Vygotsky, 1978). In particular, I have drawn extensively from Heritage (1984), Lave (1988), Suchman (1987), and the work of Schegloff (e.g., Schegloff, 1972, 1981).

Ethnomethodology takes as its object of study the everyday practices by which people create and maintain the shared world we know as "reality." Fundamental to this framework is thus the principle of social phenomena as *interactional achievements,* products of joint work by people to display to one another their developing understandings of the contexts they inhabit, and their efforts to establish a common understanding. Central to the principal of interactional achievement is the reflexivity of action, understanding, and context, such that contexts are called into being, transformed, and maintained by the very actions which appear—to traditional modes of analysis—to rest on predetermined definitions of those contexts. Contexts thus do not come prelabeled, to be thereby deterministically acted "in," but are, rather, created by the interactional work of participants, acted on and through. Language—by which we mean language in use, or discourse/conversation—in this view provides a powerful, perhaps the most powerful, set of practices for the mutual constitution of action, understanding, and context. For example, in describing the common Western greeting sequence, Heritage (1984) commented:

> The norm ["return a greeting"] is thus *doubly constitutive* of the circumstances it organizes. It provides both for the intelligibility and accountability of "continuing and developing the scene as normal" and for the visibility of other, alternative, courses of action. It follows, therefore, that whatever the outcome of the "choice," the availability of the norm will provide a means by which the conduct and its circumstances can be rendered sensible, describable and accountable. (p. 108)

Through the recognition of the overriding significance of language-in-use for the minute working of larger social structures, a distinct subfield of ethnomethodology devoted to looking at language-in-use has evolved, now commonly known as *Conversation Analysis*. Conversation Analysis unites the insights of ethnomethodology with explicit tools for analyzing everyday conversation; it is an ideal paradigm for exploring human tutoring dialogues, in that it provides a suitable theoretical foundation for understanding how tutor and student work together to construct the practices of tutoring, and it provides a collection of methods for transcribing and analyzing in fine detail the data that could enable this understanding (some of these methods are described in chapter 3).

A compatible view, with very different origins, can be found in the writings of Vygotsky (1978) and various other Soviet psychologists (e.g., 1981), as well as in the recent writings of certain western social scientists (e.g., Giddens, 1979; Lave, 1988), which I refer to here as *activity theory* (based on Lave, 1988).

In his research on child development, Vygotsky found that the child's linguistic productions and their accompanying activity were completely intertwined, each shaping and giving significance to the other:

> A child of five-and-a-half was drawing a streetcar when the point of his pencil broke. He tried, nonetheless, to finish the circle of wheel, pressing down on the pencil very hard, but nothing showed on the paper except a deep colorless line. The child muttered to himself, "It's broken," put aside the pencil, took watercolors

instead, and began drawing a *broken* streetcar after an accident, continuing to talk to himself from time to time about the change in his picture. The child's accidentally provoked egocentric utterance so manifestly affected his activity that it is impossible to mistake it for a mere by-product . . . (p. 31)

In this example, the utterance "It's broken" is provoked by the perceived context of a broken pencil, but then quickly transforms the context to one of a broken streetcar, mutating the referent of *it,* as well as the exact interpretation of broken (because a streetcar is broken in a very different way from a pencil). Thus, in much the same way that ethnomethodologists see language and context as mutually constitutive, Vygotsky treated language (speech) and activity as continually influencing one another.

These paradigms rest on assumptions that are radically different from those of more traditional approaches to both cognitive and social structures. First, these paradigms assume that cognitive processes are thoroughly socially structured, so that in a very basic sense cognition and social interaction are inseparable:

> there is reason to suspect that what we call cognition is in fact a complex social phenomenon. The point is not so much that arrangements of knowledge in the head correspond in a complicated way to the social world outside the head, but that they are socially organized in such a fashion as to be indivisible. "Cognition" observed in everyday practice is distributed—stretched over, not divided among—mind, body, activity and culturally organized settings (which include other actors). Empirical support for this proposal has emerged recently from research exploring the practice of mathematics in a variety of common settings. . . . The specificity of arithmetic practice within a situation, and discontinuities between situations, constitute a provisional basis for pursuing explanations of cognition as *a nexus of relations between the mind at work and the world in which it works.* (Lave, 1988, p. 1; italics added)

Second, both paradigms insist that all human behavior, including such apparently context-independent behavior as solving physics problems, be studied in its natural social setting:

> The AMP [Adult Math Project] investigated arithmetic practices in a variety of settings to gain a different perspective on problem solving from that found in school or laboratory. The research focused on adults in situations not customarily considered part of the academic hinterland, for no one took cooking and shopping to be school subjects or considered them relevant to educational credentials or professional success. AMP "experts" were grocery shoppers rather than physicists. . . (Lave, 1988, p. 3)

Third, they eschew rule-governed formulations of human behavior, seeing the regularity of human activity instead as achieved by choiceful actors, in contexts where meaning is created by norms that organize the interpretations of actions-taken-in-context but that do not determine those actions or their interpretations (see also chapter 9).

> In sum, any detailed investigation of the organization of social action will rapidly arrive at the conclusion that, as Garfinkel proposes, social action cannot be analysed as "governed" or "determined" by rules in any straightforward sense. This is so for two basic reasons. First, even where indisputable rules of conduct can be formulated, their relevance to action will be found to be surrounded by a mass of unstated conditions which are, in various ways, tacitly oriented to by social participants. Thus even in the simplest cases . . . an analysis of social action simply in terms of the rule will grossly understate the complexity of the scene of activity as it is available to the participants' reasoning procedures.
>
> Second, many classes of actions are not analysable by reference to clear-cut rules which either delimit them as a class or, still less, could be held to constrain or determine their empirical occurrence. Rather these actions are produced and recognized by reference to reasoning procedures which draw upon complex, tacit and inductively based arrays of "considerations" and "awareness." (Heritage, 1984, p. 128)

And fourth, as noted earlier, both paradigms see action (including language behavior) and context as mutually constitutive, that is, they provide for one another's interpretation and significance.

And what is the relevance of these theoretical principles to the study of human–human tutoring?

Obviously, because these principles apply to all forms of human interaction, they apply to human tutoring, so the first line of relevance is that human tutoring is constructed by these principles: Analyzing tutoring dialogues thus requires an eye to these informing principles.

Less obviously, they cast new light on the practice of tutoring. The current study, with its attention to the interactive roles of activity, context, and language, making use of naturally occurring tutoring sessions, may thus provide a new kind of data with correspondingly new assumptions and conclusions.

But most importantly, these principles, and the frameworks and methodologies within which they are situated, bring to life in the reality of lived time and from the perspective of the participants the recurrent methods by which teaching and learning are accomplished in a one-on-one setting, the prototypical cognitive event seen as a thoroughly social achievement.

ORGANIZATION OF THE BOOK

This volume is divided into four sections. Part I, the introductory chapters, includes chapters 1 and 2; chapter 2 presents the design of the project and the methodologies used in recording and analyzing the data. Part II, the body of the study, consists of chapters 3–7, and is organized according to the temporal structure of the tutoring sessions themselves. Chapter 3 provides an overview to the structure of the tutoring sessions; chapter 4 analyzes the openings of the tutoring sessions and the beginnings of units within each session; chapter 5 presents a description of the processes of correction and intervention in the sessions; chapter 6 explores the role of interactional timing in the diagnostic work of the tutor and student; and chapter 7 describes the closings of the tutoring sessions and the endings of units within each session. Part III, covering chapter 8, connects the face-to-face tutoring sessions with issues outside those sessions: Chapter 8 addresses the issue of mode of communication, or bandwidth, comparing the face-to-face sessions with their terminal-to-terminal counterparts; Part IV, consisting of chapters 9 and 10, provides more in-depth theoretical and concluding remarks.

2 Methodology

DATA COLLECTION

The data for the study were gathered by bringing together tutors and students from a major university. Graduate students from chemistry, physics, math, and computer science were located and asked to participate as tutors in the study. All were recommended by their departments and all had worked as tutors before. Although they all were skilled tutors, I have no information on how they were trained as tutors. They were paid for their work, in keeping with normal tutoring arrangements.

The first tutor who agreed to participate in the project was a woman, and because conversational and instructional differences between men and women are well documented (e.g., Tannen, 1990; Thorne, Kramerae, & Henley, 1983), I decided to keep gender for tutors constant. As a result, all of the tutors were women. The significance of this fact for the particular findings reported here is unclear. There has been no documented difference between men and women at the level of general conversational processes; for example, there are no differences reported in the language and gender literature for repair strategies, or for patterns for opening or closing conversations. It is also well known that there is a great deal of variation in language within each gender category (Coates & Cameron, 1988), and in fact the tutors in this project differed substantially from one another on interactional dimensions. Nonetheless, given previous findings on gender differences, and given the novelty of women tutors in science and math, the choice of women tutors has almost certainly influenced the findings of the current study. Of course, the same would be true of a study that used only male tutors.

Students were located by advertising in the student newspaper for people interested in being tutored as part of a research project. The students were both male and female. They represented a fairly wide spread of expertise, from nearly complete novice to intermediate. Their past experiences with tutoring, if any, is not known.

Initially, each student–tutor pair met for 1 hour of face-to-face tutoring in a small office, and they were video- and audiotaped. The pairs then met for two additional 1-hour sessions. In the first of these, the tutor and student were placed in different rooms and were asked to carry out their tutoring using computer terminals linked to a VAX 11/780. Each was given brief instructions about how to use the terminals and what to expect in the way of time lags, and so on; otherwise, they were given no instructions regarding the tutoring, either with regard to structure or to content. Whatever the tutor and student typed in to the computer was saved in a file; we therefore had a ready-made transcript of each of these terminal-to-terminal sessions.

The turn-taking system employed in the terminal-to-terminal condition allowed only one person at a time to type. Each person could type up to 20 lines of text per turn; at the end of each turn, the person hit the return key, his or her screen cleared, and the text was sent to the other participant. At that moment, the sender of the text lost the ability to enter text.

Upon receiving the sent text, the second person was given the ability to enter his or her own text. That person's turn ended when he or she hit return, at which moment the turn reverted to the first person.

We included this manipulation in order to examine which of the processes seen to be basic in face-to-face communication occur in keyboard-mediated communication, and how the strategies used by the participants to create these processes differed across the two modes. This condition of course does not provide direct empirical data on the nature of human–computer communication, because both participants in this condition are human. Nonetheless, from these data we can determine if basic processes such as repair can survive at all the shift to keyboard mediation, and if they can, how they are negotiated under such difficult communicative conditions. These data may thus prove relevant to the design of certain human–computer interfaces by providing potentially unexplored strategies for accomplishing basic dialogue processes.

In the last session, the student was told that he or she was to be escorted to the same terminal as in the second session; but this time he or she was told that the tutor would be a **computer** tutor rather than his or her regular human tutor. The student was given instructions again about how to use the terminal and was told that the computer tutor could handle any kind of input a human tutor could handle. In reality, however, the students in this session were communicating with the same tutor they had been working with in the previous sessions. As in the previous session, the communications between tutor and student were saved automatically by the system. The same turn-taking system was employed as in the second condition.

We included this bit of subterfuge in order to examine the effects of social expectations on tutoring; in particular we wanted to know if students would ask the same kinds of questions of computers, and use the same kinds of communicative strategies, as they would with human tutors. Given that our results on these questions are highly tentative, I have chosen not to report them here; interested readers should consult Fox (1990) for these preliminary, but suggestive, findings.

After the last session, the students were asked to fill out a questionnaire about the study, including how effective they found the computer tutor to be in comparison with the human tutors, what they liked about being tutored, and so forth. Each student was then debriefed.

TRANSCRIPTION NOTATION

Detailed transcripts of each face-to-face session were made, on the basis of both the audio- and videotapes, following the conventions of Sacks et al. (1974). In order to help the reader recreate as much as possible the sound of each fragment used in later chapters, I provide here a brief description of each of the major transcription conventions used.

A double slash (//) indicates the place at which a speaker's utterance is overlapped by talk from another speaker.

(1)
T: but I know on physics exams, .hh you have to strea:mli//ne.
S: Aha

Thus, with this notation we can see that the student's utterance starts after the *i* in the tutor's *streamline*.

An utterance that has more than one double slash in it is overlapped at more than one place, and the utterances that do the overlapping are given in sequential order after the overlapped utterance.

(2)
S: A//d them vector//ially, yeah.
T: You have no angles.
T: Right everything's on the line.

Here, T's utterance overlaps S's starting at *add* and again within *vectorially*.

A left-hand bracket at the beginning of two lines indicates that the two utterances begin simultaneously.

(3)
T: You were going to say (2.7) charge on the electron>
    (0.2)
T: Right?
   [
S: Aha

Tutor and student begin talking simultaneously.

The equal sign (=) indicates latching, that is, the next speaker begins without the usual "beat" of silence after the current speaker finishes talking. In this case there is an equal sign at the end of the current speaker's utterance and another equal sign at the beginning of the next speaker's utterance. If two speakers simultaneously latch onto a preceding utterance (i.e., they begin talking simultaneously), this is indicated in the transcript with a left-hand bracket preceded by an equal sign.

(4)
T: You've got meters, kilograms, meters, seconds.=
S: Seconds.
   =[
T: So this is in ess ay units.

Here S and T simultaneously latch onto T's utterance.

Numbers given in parentheses indicate elapsed silence, measured in tenths of seconds. Single parentheses with a dot between them represent a silence that is less than one tenth of a second but still longer than the usual beat of silence.

Certain facts about the production of the talk are given through the orthographic symbols used. Punctuation is used to suggest intonation; italics indicates stress. A colon after a letter means that the sound represented by that letter is somewhat lengthened; a series of colons means that the sound is increasingly lengthened.

The letter h within parentheses indicates "explosive aspiration," and usually means some type of laughter is being produced. A series of hs preceded by a dot represents an inbreath (where number of hs is meant to correspond to the length of the inbreath), whereas the same series preceded by nothing represents exhaling.

Questionable transcriptions are enclosed within single parentheses; the transcribers thereby indicate that the exact form of the utterance is not clear. A speaker's initials given in single parentheses means that there is some question about the speaker's identity. Single parentheses with capitalized words enclosed

—for example, (CLEARS THROAT)—represent nontranscribed material (i.e., noise that is nonlinguistic).

">" indicates rising, but not terminal rising, intonation, as is often found at the end of each member of a list.

Readers interested in the theoretical underpinnings of this transcription technique should see Sacks et al. (1974), Heritage (1984), and Ochs (1979).

3
Tutoring Dialogue Structure

ORGANIZATION OF EVERYDAY CONVERSATION

Because tutoring dialogues can be viewed as a somewhat specialized form of everyday conversation, and because they are constructed on the basis of the norms (Sacks et al., 1974) that construct everyday conversation, I have chosen to use terminology and methods of analysis originally developed for the analysis of everyday conversation.

The major components of the methods and practices that are used both to analyze everyday conversation and to construct it in the first place are: *turns*, produced in general by one person; and what are called *adjacency pairs*, produced in general by two different parties. With these two sets of notions, we can describe much of what happens in conversation.[1]

The Turn-Taking System

One of the most critical aspects of conversational structure is turn-taking (Sacks et al., 1974). How is the orderliness of participants speaking (for the most part) one at a time achieved?

We need first to establish that there are units out of which turns can be constructed. These units have been referred to as *turn constructional units* (TCUs), and can be single lexical items (*yes*), phrases (*in the box*), clauses, or sentences. According to the turn-taking system, each speaker can claim one of

---

[1] For a more complete introduction to Conversation Analysis, see Levinson (1983), Atkinson and Drew (1979), and Heritage (1984).

these TCUs. The end of such a unit constitutes a place where speaker change could occur; that is, at this point another person could begin talking. The end of a TCU is thus a *transition relevance place* (TRP), because it is a place at which a transition from one speaker to another can (but need not) occur.

Within the Conversation Analysis tradition, the normative system of turn-taking has been described by means of a set of turn-taking rules (Sacks et al., 1974). Although they are formulated in the terminology of rules, it should be understood that they are not rules in the traditional linguistic or computer science sense; they are norms that speakers orient to rather than determinors of behavior. Their nondeterministic status can be seen in the fact that speakers can violate them (e.g., one person can begin speaking before the other has come to a possible completion point), and these violations are given interactional interpretations (e.g., the person is interrupting and being rude).

The following turn-taking rules, which are based on these concepts, are taken verbatim from Levinson (1983), which is based on Sacks et al. (1974):

Rule 1 applies initially at the first TRP of any turn.

(a) If the current speaker selects a next speaker in current turn, then current speaker must stop speaking, and that next speaker must speak next, transition occurring at the first TRP after next-speaker selection.
(b) If current speaker does not select next speaker, then any (other) party may self-select, first speaker gaining rights to the next turn.
(c) If current speaker has not selected a next speaker, and no other party self-selects under option (b), then current speaker may (but need not) continue.

Rule 2 applies at all subsequent TRPs.

When Rule 1(c) has been applied by the current speaker, then at the next TRP Rules 1(a)–(c) apply, and recursively at the next TRP, until speaker change is effected.

These rules provide a foundation for making sense out of two related phenomena: simultaneous talk and silence.

Simultaneous talk obviously occurs when two (or more) speakers talk at once. But not all simultaneous talk represents a violation of the turn-taking system rules. Rather, there are two basic types of simultaneous talk (i.e., overlap): competitive overlap and noncompetitive overlap. In one type of noncompetitive overlap, called *terminal overlap,* the current speaker approaches the end of a turn constructional unit, and, as that is happening, the next speaker, having predicted the type of TCU that the current speaker is producing, starts talking, thus overlapping with the very end of the TCU that the current speaker was heard to be constructing. This type of overlap is not heard as competitive. In addition, laughter from one party simultaneous with talk from another party is often not heard as competitive (but appreciative). An example of terminal (noncompetitive) overlap is given here:

(1)
(HG:II:4-5)
N. Also he sid that (0.3) 't what you ea:t, (0.2) end how you wash yer face has nothing tih do with it,
   (0.8)
H. Yer kiddin//g.
⇒ N. nNo:,

In *competitive overlap*, on the other hand, the rules of the turn-taking system are violated, usually by the next speaker starting up before the projected transition relevance place of the current speaker's TCU. An example of competitive overlap follows:

(2)
(HG:II:8)
H. En I nearly wen'chhrazy cz I//: l:lo:ve that mo:vie.
⇒ N. y:Yeah I know you lo:ve tha::t.

In this passage, H is not near the end of a TCU (she has just gotten out the subject of a subordinate clause) when N starts up.

Competitively overlapping utterances can be characterized by higher pitch, slower tempo, louder volume, and lengthened vowels.

Silence occurs, obviously, when no one is talking. Not all silences are equivalent, however. Silence is considered a *pause* if it is attributable, by the turn-taking system, to a given party; for example, if current speaker has selected next speaker, then any silence after current speaker reaches the end of his or her TCU is a pause attributable to the selected next speaker. The following is an example of two pauses, both attributable to Speaker B (I am following Levinson, 1983, in this definition of gaps and pauses, the latter of which he called "attributable silences"):

(3)
(from Levinson (1983))
A. Is there something bothering you or not?
⇒   (1.0)
A. Yes or no
⇒   (1.5)
A. Eh?
B. No.

Silence is considered a *gap*, on the other hand, if it is not attributable, by the turn-taking system, to any particular party. This situation often arises if the current speaker has not selected a next speaker, and the silence therefore "belongs" to no one (although the current speaker can apply Rule 1(c) and get another turn at talk, and in certain cases this will create the effect that the preceding silence was in fact attributable to the current speaker). An example of a gap is given here:

(4)
(HG:II:25)
N. =.hhh Dz he av his own apa:rt//mint?
H. .hhhh Yea:h,=
N. =Oh:,
⇒   (1.0)
N. How didju git his number

The turn-taking system proposed by Sacks et al. (1974) affords insights into many other aspects of conversation, but for our purposes here the concepts covered previously suffice.

The Structural Organization of Conversation

Another level of organization is the adjacency pair.

An adjacency pair is said to consist of two parts, the first of which (the first-pair-part) makes relevant a particular type of action (the second-pair-part) from another party. In fact, it is not altogether clear that there should only be two parts in a "pair"; nor is it clear that the parts need to be adjacent. These problems with the literal interpretation of the term *adjacency pair* need not concern us here.

We thus have first actions that make relevant second actions by another party. Standard examples of the notion of adjacency pair are question–answer, invitation–acceptance, offer–acceptance, request–comply, announcement–assessment, and so on.

An adjacency pair, or sequence, can take three types of expansions: preexpansions, insert expansions, and postexpansions (see Levinson, 1983, for a discussion). A preexpansion is an adjacency pair that comes before another adjacency pair and is preliminary to it. A classic example of a preexpansion is the preannouncement adjacency pair, which usually consists of a preannouncement first-pair-part and a clearance second-pair-part:

(5)
(invented example)
A. Guess what?   [Pre-announcement]
B. What?   [Clearance]
A. I got an IBM PC!   [Announcement]
B. That's great!   [Assessment]

An insert expansion is an adjacency pair that comes between the first-pair-part and the second-pair-part of another adjacency pair. Question–answer pairs are common insert expansions:

(6)
(invented example)
A. May I speak to Arnold?   [Request]
B. May I ask who's calling?   [Question]
A. Nancy.   [Answer]
B. Ok.   [Comply]

Notice that the insert expansion pair is completed before the other pair continues.

A postexpansion is a pair that follows another pair. If, for example, an adjacency pair that seeks to repair some source of trouble in a preceding turn (known as a repair sequence) is initiated after the possible completion of an adjacency pair, the repair sequence will be considered a postexpansion:

(7)
(invented example)
A. Do you like Virginia?   [Question]
B. Yeah.   [Answer]
A. You do?   [Next turn repair
                              initiator, pre-disagreement]
B. Well, not really.   [Repair, backdown]

It is worth taking a brief excursion at this point into the nature of repair sequences, because they are treated in greater depth in chapters 5 and 8. A repair sequence is often initiated with what is called a *next turn repair initiator* (NTRI), which indicates that in the next turn the next speaker should attend to some problem that the current speaker has encountered with the preceding turn; hence it initiates repair action for the next turn.

(8)
[invented example]
A: You should eat all your spinach.
B: What?   [NTRI, trouble in
                              previous utterance]
A: Eat your spinach.   [repair]

Such insert expansions can be done recursively, such that an inserted sequence can have within it another inserted sequence:

(9)
[invented example]
A: You should eat all your spinach.
B: Eat all my what? [NTRI]
A: Hmm? [NTRI]
B: What should I eat all of? [repair, NTRI]
A: Your spinach. [repair]

A next turn repair initiator does not only indicate technical difficulties with hearing, and so forth, however; it often indicates that the speaker of the NTRI may do a disagreement with the turn flagged for repair.

Additional material about this method of analysis can be found in Atkinson and Drew (1979), Goodwin (1981), Heritage (1984), Schegloff, Jefferson and Sacks (1977), Levinson (1983), Pomerantz (1975), Schegloff and Sacks (1973), Schenkein (1978), and Terasaki (1976).

ORGANIZATION OF TUTORING DIALOGUES

Given these fundamental notions of conversational interaction, we can now turn to exploring the organization of the tutoring sessions. This section provides a rough overview of the basic structural properties of the sessions; details of their organization can be found in chapters 4–7.

The sessions were for the most part organized around solving particular problems (one session was much more directly instructional than problem solving; see section on instruction).

Problem Solving

As we see in chapter 4, solving individual problems leads to a question–answer organization at the highest level of structure, that is, the problem statement implicitly or explicitly asks a question that in the ideal case will be answered by the student; but because the question in fact projects a sequence of actions rather than just an immediate answer (as is usually the case in everyday question–answer pairs), there is a great deal of complex interaction that takes place between the initial question posed by the problem statement and the final answer.

As we see in chapter 4, the solving of a particular problem often starts with the tutor or the student reading the problem statement out loud:

(10)
S: It's saying (READING) if secant of theta, (0.6) okay> now, secant of theta (1.7) equals three. And theta is a fourth quadrant angle, find tangent of theta.

(11)
S: Number seven is the one I was going // to do.
T: (READING) How close must two electrons be if the electric force between them is equal to the weight, (0.7) of either at the earth's surface.

The tutor or student may embed within the reading of the problem statement question–answer sequences:

(12)
⇒ T: (READING) A one ohm wire. Okay, what's ohms
    (1.4)
⇒ S: It's resistance.
T: Okay. Is drawn out to three times its original length. What is the resistance now.

(13)
S: A:nd, in this case they say (0.3) co:secant (0.8)
⇒   tee? Theta?
    (1.3)
⇒ T: Yeah. They don't care. Whatever.
   [
S: (Is that- is that)
S: Yeah, okay.

After the problem is read, the student may describe how he or she went about solving the problem, in a kind of narrative form:

(14)
S: I'm going to say the secant of th- uh, theta, .hh the secant is rea:lly uhm (0.5) one over (a) cosine (0.4) theta, right?
S: So it's a reciprocal.
T: Mh//m
S: Identity. Okay? .hh So I have a secant (0.9) theta equal one over cosine theta. .hh Okay, it says- find the tangent of theta.

Notice that embedded within this narrative are requests for confirmation (*right?*) and confirmations, in the form of question–answer sequences, which are a kind of repair sequence. This kind of embedding is extremely common.

Embedded within the student's verbal working of a problem, we also find checks for understanding from the tutor, often in the form of question–answer sequences but sometimes in the form of other adjacency pairs:

(15)
S: The way I understand it is say (2.1) say you have two charges here, we'll call this one the big q and this one the little one. A//nd the q
T: **Now let's just look at this one.**
S: **Right.**

In some cases, after the problem is read the tutor begins a sequence with a question about how the student proceeded or how the student could proceed:

(16)
T: (READING) How close must two electrons be if the electric force between them is equal to the weight, (0.7) of either at the earth's surface.
    (0.3)
⇒ T: .hh Okay, so what did you:
    (0.5)
S: So this is what- we're allowed to have our little um, (0.4) sheet of formulas

In answering such a question, the student may engage in a narrative of how the problem was solved in the past or is being solved now, which can then take insertion sequences.

That insertion can be many layers deep can be seen from the following passage. I have indented the text where a new level of embedding occurs; by this analysis there are at least four different levels here, and this is by no means an unusually elaborate interchange:

(17)
T: (READING) What is the speed of a 350 ev electron. Okay.

    T: And what are these, these are> (0.9) those aren't lengths, so what are they
    S: That's the work?
        T: Work or e//nergy.
        S: Energy>
⇒   T: Okay? So this is an energy.
    T: And then it's as- it- it's actually
    S: Oh I see.
    T: So they want you to relate speed to energy.
    S: Oh, okay.

(0.9)
T: Okay? So what (1.5) what- equation do *you* know, that is an energy equation, that has the velocity of (0.7) the particle in it.
S: Oh boy, it's been such a long time.
(1.6)
T: But th*a*t's the answer.
S: Yeah.
(1.1)
S: They expect us to remember things from // ( )
T: That's an easy one, it's an easy one // ( )
S: It's an easy? Okay. So I I need this is an energy and I need to find energy related to speed?
T: Yeah.

Problem solving goes on in this way, with the student narrating steps, the tutor asking questions or making suggestions, the student asking for confirmation, the tutor checking understanding, and so on, in some cases with multiple levels of embedding, until the tutor and student agree they have come to an acceptable stopping point (see chapter 7 for a discussion of this part of the process).

Instruction

One of the sessions is organized less in this problem-solving structure than the others, being more engaged with instruction of concepts. The structure of most of this session (and parts of the other sessions) is more like a habitual narrative, which details the sequence of actions which lie behind the object descriptions of concept labels (see chapter 4 for a similar discussion).[2]

(18)
T: Okay, so (1.1) chain rule?
(1.5)
T: Ring a bell?
S: Yeah, yeah // chain rule rings a bell.

---

[2]Garfinkel described the relationship between object descriptions and actions:

> . . . an actor's treatment of a description will unavoidably address it as *contexted,* as unavoidably an *action* which maintains, transforms, or, more generally, *elaborates* its context of occurrence and, hence, as unavoidably a *temporally situated phase of a socially organised activity.* (Cited in Heritage, 1984, p. 156)

The tutors, as part of their work of contextualizing abstract concepts, formulas, and problem statements for students, thus provide elaborate mechanisms for students for "converting" object labels (like *chain rule*) into action in particular contexts.

T: Okay.
T: Okay. So what that says is if you have (2.1) a function sitting inside of another function.
   (0.8)
S: Right
T: (And) to diff*e*rentiate it, you take the *out*side derivative (1.0) the ef prime (1.7) and then you multiply it by the inside derivative, (0.6) the gee prime.

Within this habitual narrative—habitual in the sense that one "always" carries out these actions under the appropriate circumstances—repair insertion sequences can occur, as can other kinds of insertion sequences. This is similar to the kinds of repair sequences we saw in the section on problem solving.

(19)
T: Now, with the natural log as it *is*, the only numbers you can put in, are positive numbers. And, if you have the absolute value there, now you can put negative numbers in for *u,* when you take the absolute value, it turns it into the positive number, and the natural log will accept it. Okay.
S: **(BREATHES IN LOUDLY)**
T: **Go ahead.**
S: **Let's ahyee digress for a second here**
T: **Okay**
S: **Someth- I've always had a problem with absolute value.**
T: **Mkay.**

One also finds postexpansion sequences, in which, for example, the student or tutor initiates a clarification question–answer sequence after the tutor has come to a possible completion point for the narrative:

(20)
T: Rational functions were (0.7) real nasty.
S: Yeah.
T: **So Mkay, would you like to look at more examples of this.**
S: **Yeah. That would be good.**

Narrative structures like these are different from question–answer structures like the ones we saw before in that they provide different opportunities for speaking for the two parties, to wit: A narrative is produced essentially by one person, with opportunities for the other person to speak consisting of (a) finishing an utterance with or for the producer of the narrative (see chapter 6 for a detailed discussion of this phenomenon); (b) insertion sequences, like repair sequences; and (c) postexpansion sequences, like questions for further elaboration on a point. A question–answer sequence, on the other hand, is in its essence produced by two different people (even though there are occasions—see chapter 5—in which the same person produces both question and answer). The problem-solving question–answer structure thus in some sense "requires" student involvement in a way that the narrative structure does not.[3]

None of these structures are unusual for everyday face-to-face interaction. The units are the same as everyday conversation, and they are constructed apparently by the same kinds of principles that operate in everyday conversation. Perhaps the distinguishing characteristics (with regard to everyday conversation) are (a) their engagement in a task that requires building a series of concrete actions on the basis of a description of objects and relations (although similar situations can arise in other activities, as for example in following a recipe together); and (b) the asymmetry in the dialogue between the tutor and the student (although, again, similar situations arise between parents and children). The first point is discussed in detail in chapter 4; given the importance of the second point, I turn now to an exploration of the effects of tutor–student asymmetry on the organization of the tutoring sessions.

ASYMMETRY BETWEEN TUTOR AND STUDENT

In some institutionalized contexts (e.g., the courtroom), conversational roles and responsibilities are carefully prescribed, such that one party is designated as, for example, questioner and the other as answerer, and violations of these heavily constrained roles can lead to severe sanctions (e.g., being ruled in contempt of court). The conversational asymmetries among various participants in such settings are obvious even to a casual observer.

The asymmetries in our tutoring sessions, in contrast, are neither prescribed by a body of rules nor entirely obvious. Both participants ask and answer questions, initiate repair, work through steps in solving problems, make calculations, explain concepts, make mistakes, and so forth, all negotiated at the moment of utterance. Nonetheless, there are regular and recurrent differences in the conversational behavior of tutors and students that manifest themselves, at least in part, through differences in dialogue structure.

The basic asymmetry between tutor and student can be characterized in terms of the functions of tutoring: If, as I have claimed, the fundamental function of math and science tutoring is the contextualization by the tutor for the student of abstract object descriptions into concrete actions on concrete objects, then it

---

[3]This statement does not imply that narratives are noninteractive. See Schegloff (1981) for a cogent discussion of the pervasively interactive nature of narrative.

follows that the chief task of the tutor is to learn how this particular student, in this particular setting, can best situate these particular problem statements; then, the chief task of the student is to learn general practices of contextualization from working with particular problems. In other words, as one would expect, the tutor is focused on learning about the student, and the student's relationship to the domain under study, whereas the student is focused on learning about problem solving within the domain under study. Of course, the tutor may actually learn something about the domain, but this is in some sense a secondary gain; what is crucial for the tutor to learn about is how to enable this student to contextualize problem statements.

As an important aspect of these different tasks, both tutor and student act on the assumption that the tutor knows the domain "correctly" (or at least more correctly than the student), whereas the student knows the domain only partially and in many ways "incorrectly."

Several conversational differences emerge from these basic asymmetries in task and assumed knowledge. Although students and tutors both ask questions, student questions are in general interpreted as requests for clarification, requests for help with figuring out the next step, and so forth, whereas tutor questions are in general interpreted as seeking information about the student's understanding. In the following pair of fragments, for example, the two questions look fairly similar: but the tutor's question seeks to understand how the student proceeded on the problem, and thus how the student goes about contextualizing problem statements; the student's question, on the other hand, is a repair initiator that seeks help from the tutor in completing a particular calculation that the tutor showed the student a bit earlier:

(21)
[a]
T: (READING) H*ow* close must two electrons be if the electric force between them is *e*qual to the w*ei*ght, (0.7) of *ei*ther at the earth's surface.
   (0.3)
⇒ T: Okay, **so what did you:**
[b]
S: Minus I guess the x component of this one. (1.2) So (0.9) oops the y components (1.1) so it'll be (2.1) uhm (11.0) Instead of finding theta I'm
⇒   going to (0.5) do it by uhm (2.3) **oh, how'd you do that**

Although the two questions are grammatically similar, they are heard as requesting very different kinds of responses, and although they both engender question–answer adjacency pairs, 21(a) prompts an answer that could be seen as encompassing all of their work on that problem. On the other hand, 21(b) prompts a fairly brief side sequence in which the tutor goes through the steps of the relevant calculation. The asymmetry affects the interpretation of utterances that tutor and student produce.

Nevertheless, tutors can initiate short repair or clarification sequences with questions:

(22)
((Student working problem: 10.8))
⇒ T: **Where did this minus sign come from.**
S: It's minus ex. This minus ex shouldn't be here.

Although asking a question is not the most common strategy for the tutor to initiate repair, it does occur. But even here the asymmetry appears: in 21(b) the student's repair initiator does not initiate correction but rather a kind of repetition; in (22), however, the tutor's repair initiator initiates correction and not merely repetition. Again, even though tutor and student behavior appears on the surface to be structurally identical (question–answer sequence, or even NTRI–repair sequence), they in fact involve quite different actions: repetition versus correction.[4]

Moreover, when tutors offer explanations or work through steps of a problem out loud, their utterances are heard in general as correctly describing how to proceed on a problem or how to define a concept, and the student's task is to learn about how to contextualize problem statements from them. When students offer explanations or work through steps of a problem out loud, however, it is so that the tutor can "inspect" the student's understanding and initiate repair if necessary. The conversational operations that a student can perform on a tutor's explanation are correspondingly different from those a tutor can perform on a student's proffered explanation: A student can ask for clarification and elaboration, and show understanding, whereas a tutor can initiate repair, correct, show agreement, and redirect the line of explanation. Although their behaviors may be superficially similar (e.g., both a tutor and a student can say *right* in response to an offered explanation), the asymmetry of tutor and student means that the "deep" organization of their dialogue is really quite different. For example, roughly speaking, when a tutor says *right* after a student's explanation, it means "your explanation is correct," whereas when a student says *right* after a tutor's explanation, it means "I understand." Searle's (1979) shopping-list analogy offers some relevant insights:

> Suppose a man goes to the supermarket with a shopping list given him by his wife on which are written the words "beans, butter, bacon, and bread." Suppose he goes around with his shopping cart selecting these items, he is followed by a detective

---

[4]In some cases, of course, the tutor asks for clarification and the student initiates correction of tutor behavior. But in general the asymmetry holds.

who writes down everything he takes. As they emerge from the store both the shopper and detective will have identical lists. But the function of the two lists will be quite different. In the case of the shopper's list, the purpose of the list is, so to speak, to get the world to match the words; the man is supposed to make his actions fit the list. In the case of the detective, the purpose of the list is to make the words match the world; the man is supposed to make the list fit the actions of the shopper. This can be further demonstrated by observing the role of "mistake" in the two cases. If the detective goes home and suddenly realizes that the man bought pork chops instead of bacon, he can simply erase the word "bacon" and write "pork chops." But if the shopper gets home and his wife points out that he has bought pork chops when he should have bought bacon he cannot correct the mistake by erasing "bacon" from the list and writing "pork chops." (p. 4)

In the same way, although a tutor and student may produce what appear to be the "same" kinds of explanation, like the two lists in Searle's example they bear different relationships to the ongoing activity, including the structural opportunities they create for future dialogue.

The interpretation of silence, too, is constructed within this asymmetrical framework. Although a student's silence is often "heard" as indicating confusion (unless the student is displaying signs of working successfully on a problem), a tutor's silence is often treated as indicating upcoming repair or correction:

(23)
T: Which (in math) is generally lit- written two root two.
⇒   (0.8)
S: Two root two. Okay. Hold on just a sec. Two root two, (0.4) Is that like (2.2) so?
⇒   (0.4)
T: Just- two times.

In (23), the (0.8) bit of silence presages the student's confusion about the phrase "two root two," whereas the (0.4) bit of silence (after the student's understanding check) suggests the tutor's upcoming correction of the student's understanding.

There is one last set of differences in conversational organization that arises from the asymmetry between tutor and student, namely, the distribution of responsibility for various segments of the dialogue. Although all of each dialogue is mutually negotiated between tutor and student, requiring constant input from both participants, it appears that at different points in the dialogue one participant may shoulder ultimate responsibility for guiding or initiating the sequence. For example, as we see in chapter 4, inasmuch as it is the student's trouble with a particular domain that represents the occasion for the tutoring, the student has ultimate responsibility for describing that trouble so that it can be used as a starting point for further negotiations about how to proceed with the tutoring.

That is, whether the student offers the information without being prompted or only provides the information after being prompted by the tutor (by a question like "where shall we start?"), it is in the end the student who must provide this crucial piece of information.

In a similar vein, although the end of each problem and the end of each session is thoroughly mutually negotiated, it seems that it is the tutor who has ultimate responsibility for shaping what is going to count as the answer, or a reasonable place to stop the session (see chapter 7 for more detailed discussion of closings). An example from one of the sessions is given here:

(24)
- T: Now I want to see something. They got a different answer. Because they solved for it in a different way.
    (0.9)
- S: Mmm.
    [
- T: Everything that *you* did was right.
- S: Mhm mhm
    (2.7)
- S: Mmm
- T: So they-
    (0.9)
- S: Mhm
- T: Uh this answer's one point three four. (0.8) Your answer in (that) was (4.6) one point one two. So it's *not* (0.9) the same.
- S: Mhm mkay
- T: Now,
    [
- S: Do you want me to work it again?
    (0.7) [tutor doesn't answer]
- S: Mkay>
    (1.8)
- T: But- what did we just solve for?
.
.
- T: Sorry. They have it- they've solved for it a lot earlier than that.
.
- T: So you're right.

S: Ah okay.
   (1.0)
S: Good.
T: Mkay.

Here the tutor checks the apparent discrepancy between the student's answer (which appears to be correct) and the book's answer (which is also presumably correct). She will not let the student go on to the next problem until she has worked through the source of the discrepancy. This is the kind of responsibility that tutors exercise both at the ends of problems and at the ends of sessions.

This chapter has covered some of the fundamental notions of dialogue organization. In the next chapter, we see how these notions are brought into play in beginning a tutoring session, and beginning a segment within a session.

4 Openings

AN OPENING ON OPENINGS

This chapter explores the openings of our tutoring sessions, as well as the beginnings of sequences of each session. The openings of sessions and segments afford a particularly good view of one of the main functions of tutoring—the collaborative contextualization of abstract forms.

I define *opening* of a session to be that portion of a tutoring session that follows the turning on of the camera and the departure of the researcher from the room and precedes the actual problem-solving or instruction activity that the tutoring session is mainly constructed to enable. I define *beginning* of a sequence to be that portion of a sequence—where the sequence itself is often bounded by adjacency pair parts (such as question–answer)—that sets up the activity projected by the sequence (e.g., solving a particular problem or going over a new concept). Openings of sessions typically have clearly identifiable boundaries; it is usually fairly straightforward to determine where the negotiations for the session stop and the problem-solving or instructional activity starts. The term *beginning of sequence,* as I am using it here, is somewhat fuzzier; the beginning of a problem-solving sequence, for example, is often neither structurally nor substantively separable from the rest of the sequence. Nonetheless, there are recurrent, and crucial, tasks involved for the tutor and student in starting in on a new sequence, as we see later, and it is the working through of these tasks, however these are conversationally structured, that draws our attention to this activity.

Although openings of sessions and beginnings of sequences are quite different analytic objects, I have chosen to discuss them together because they share what for my concerns is a critical feature: They both orient the participants (here, tutor and student) to future actions and hence illustrate many of the same fundamental issues that illuminate this study.

The chapter first describes how tutors and students negotiate how they are going to proceed, that is, what might be considered plan construction; we then move on to examine how they begin work on particular problems.

OPENINGS OF SESSIONS

In all of the face-to-face sessions we observed, the first few minutes were spent negotiating how the tutor and student would begin working together. Neither the tutor nor the student came to the session with fixed plans for how the session should proceed; and so, in some cases either the tutor or the student (or both) brought resource materials (e.g., textbooks, or sample exams), or some general ideas of what kinds of activities they wanted to engage in. In no case did any of the participants say anything like "here's what we're going to do today." A representative opening from one of the sessions is given here:

(1)
S: Mkay, so let's see now. (0.7) Ahm, maybe the best approach would be to-they like to give examples. And solutions and since I'//m so rough on this, maybe we should do something that elementary> Does that sound good?Or
T: Mhm
T: Sure.
S: Or, or if *you* have an*o*ther idea, that's ah fine by me too, I'm a novice at this, so
    (1.2)
T: Ahm, (0.5) whatever you think you need help with

It is interesting to note that all of the negotiations for how to proceed, with the exception of the session excerpted in (6),[1] are initiated by the student. Just as in the case of phone calls it is common practice for the caller to state early on in the call the reason for the call, in these tutoring sessions, after the tutor and student have identified themselves and responded to the requests from the researcher, it seemed to be the role of the student to state the trouble that brought him or her to seek tutoring—in other words, why the student has summoned the tutor (see Schegloff, 1973, for a discussion of phone calls as beginning with a summons—

---

[1] It seems clear from the tutor's question in (6) ("but you're mostly interested in starting in with integrals?") that the tutor and the student have already talked about the student's reason for seeking tutoring before the filming started. So (6) is not a counterexample to the general finding.

answer sequence). Because most universities view tutoring as remedial activity, necessary only for students who are not doing well in class, "summoning" a tutor carries with it an implicit acknowledgment of trouble, which it is now the job of the student to make explicit for the tutor. The statement of the student's trouble is the beginning of the negotiation of how to proceed, and often includes the student's own suggestion for how to proceed.

(2)
S: I'm taking math 101, fundamentals, fundamentals of algebra. And uh (2.1)
S: Just want to go over my test from last week. We've been doing uh (2.0) radicals and exponents.
T: Yeah.
S: And (3.0) let's see (TURNING PAPERS: 2.0) Got a miserable 81.

His suggestion for how to start the session is to "go over" his test, and his trouble (which, interestingly, he doesn't produce until *after* his suggestion for how to proceed, reversing the canonical order) is that he got a "miserable 81" on the test.

In the case of phone calls, even if the reason for the call rests with the caller, knowing this the receiver of the call can initiate the "reason for the call" sequence by asking, for example, "So what can I do for you." In the same way, the tutor, knowing that the reason for the tutoring session rests with the student, can initiate this part of the opening:

(3)
⇒ T: So what are you doing?
S: A//hm
⇒ T: You're taking physics.
S: Taking physics 302. I didn't have (0.4) I had 301, but I had it a long time ago at Arizona State>
T: Aha
S: So I'm kind of like (0.6) trying to regroup and // get back into it
T: ((laugh))
S: It's really had in f*i*ve weeks too.

S: So if I just *do* it, and all I'll do is physics and then (0.4) hopefully ((laugh)) (0.6) get something out of it, but (0.7) I've a t*e*st tomorrow, and I don't know if you're familiar with the book or anything, I'm sure you're familiar with the st*u*ff

S: And uhm (0.5) I think I'm understanding the concepts pretty well (1.0) uhm (0.6) a couple of them I don't really get

In talking about the trouble that brought him or her to get tutoring, the student often initiates the next part of the opening sequence, which negotiates a candidate solution to the student's trouble; in other words, the two begin to work out how to proceed with the instructional activity:

(4)
S: I think I'm understanding the concepts pretty well (1.0) Uhm (0.6) a couple of them I don't really get about (1.3) the difference between a- a battery (0.7) the (works that'll) charge it and the works that (1.1) that'll charge a mm, capacitor> And stuff like that. The formulas are //kind of similar but there's like a half in front of one of them and-
T: Aha
    (1.0)
T: Oh
S: I got a little confused on that.
    (1.1)
S: But uh=
T: =Ts yea, well let's go to the place in the book.
S: Okay.
T: ts cause uh (0.6) this is a very uh specific thing like (I) say, I don't want to say anything that//'ll mess you up. But
S: Right.
T: where- (find it) in here and let's read // like the paragraph.
S: Okay.
    (1.5)
T: Then the other thing is I usually find for studying for tests the best thing to do is just to- just pick problems, //and work them.
S: Yeah, that's what I've been d//oing. (More or less)
T: And then any concept in the problem we can like go off and talk about it for a b//it, then go back and work some.
S: Aha
S: Okay, well we could do it from that angle then

In this passage, the student first offers one of her sources of trouble, which, by the conventions of everyday conversational practice, makes relevant an offer of a solution to the trouble. Subsequently, the tutor provides a possible solution by suggesting that they look at that place in the book where the formulas for capacitance are given. They then start paging through the student's textbook, looking for the relevant paragraph. The student's self-diagnosed trouble is, thus, used as a starting point from which the tutoring can proceed. But in the process of looking for the appropriate paragraph, the tutor suggests another tutoring strategy, which involves solving particular problems and working with conceptual issues within the context of particular problems. The student accepts this suggestion and they begin working on solving problems.

As can be clearly seen from (4), the student's self-diagnosis is not taken as determining the course of the tutoring, but is treated, rather, as a starting point from which the two can pursue further negotiation. Although the student's self-diagnosis is treated by both parties as important information for establishing the course of the session, it is by no means determinative of that course.

Moreover, as Suchman (1987) made clear, whatever plans the tutor and student might construct together in the opening of the session are themselves not determinative of the course of the action of the tutoring session, but are rather resources that the participants can make use of in a variety of ways (e.g., reconstructing the sense of what they have been doing).

> The student of purposeful action on this view [i.e., the view that plans determine courses of action] need know only the predisposition of the actor and the alternative courses that are available in order to predict the action's course. The action's course is just the playing out of these antecedent factors, knowable in advance of, and standing in a determinate relationship to, the action itself.
>
> The alternative view is that plans are resources for situated action, but do not in any strong sense determine its course. (pp. 51–52)

A good illustration of this fact comes from the session excerpted in (4); after they have negotiated a problem-solving approach to the session, the tutor occasionally brings up general problem-solving "tricks" that are not directly related to the specific problems they have been solving; but when she does this she marks these comments as "misplaced" with regard to the developing structure of the dialogue. Although the tutor is not deterministically constrained by the "plan" she and the student have evolved, she does orient her behavior to that plan, and to the structure of the prior conversation, which is also oriented to the plan:

(5)
T: Okay. Now, another concept (that)'s very useful on exams, is conservation of energy.
    (0.4)
S: Aha
T: And we haven't seen a problem like this yet
    [explains conservation of energy]
T. And that's just something- I just wanted to make sure I'd mentioned (it)

The tutor indicates with her last turn that this sequence is out of place with regard to the solving of particular problems that they have been engaged in, thereby displaying her orientation to their plan and the emerging structure of the dialogue, but she nonetheless is not bound by that plan to the extent that it determines what she talks about.

As the tutors and students work together to arrive at an approach to the session, the tutors variously offer suggestions for how to proceed, modify, transform, or accept suggestions from the student. A rough analysis of the negotiation looks something like this: If the tutor accepts the student's approach, then the tutor and student set about getting "to work"; if the tutor modifies or transforms the student's suggestion, or makes the original suggestion, then the student can accept, modify, transform, or reject the tutor's contribution. If the student modifies or transforms the tutor's suggestion, the process recycles, with the tutor accepting or modifying the student's emendation, and the student can now either accept or modify the tutor's contribution. This process could in principle continue indefinitely, but in none of our sessions did we see more than a few rounds.

Although such a view of the process is accurate to a certain extent, it hides from view the entirely collaborative nature of the negotiation; the process is so thoroughly collaborative that in many instances it is extremely difficult, if not impossible, to assign a suggestion to just one of the participants. As Goodwin (1979) made clear, a sentence that is physically uttered by only one person is nonetheless thoroughly interactionally constructed:

> The sentence eventually produced emerges as the product of a dynamic process of interaction between speaker and hearer as they mutually construct the turn at talk. The fact that a single coherent sentence emerges, and that this was apparently the sentence being constructed all along, is among the more striking features of this process . . . the sentence actually produced within a particular turn at talk is determined by a process of interaction between speaker and hearer. (p. 112)

In the following passage, for example, the tutor's utterance "or do you want to review some other stuff first" is clearly made in response to the (student's) silence following her initial question and therefore the utterance is both hers and the student's, inasmuch as it is shaped by the student's silence. The student's silence shapes the tutor's utterance, and the silence is shaped by the tutor's utterance—whether it is in fact the student's silence rather than the tutor's or nobody's is in part created by what her utterance does, as is the significance of that silence (for similar phenomena see Fox, 1987a; Schegloff, 1981; Suchman, 1987):

(6)
T: So it does fractions (1.3) but you're mostly interested in starting in with integrals?
⇒ (2.0)

OPENINGS   37

T: or do you want to review some // other stuff first?
S: um
S: Yeah, let's I think we could probably do a fairly quick review ah on in differentiation

Although the 2-second silence may be a kind of rejection (or pre-rejection; see Levinson, 1983; Pomerantz, 1975) on the part of the student of the tutor's candidate understanding of how the student wants to proceed (itself a fully collaborative product), it is more open to interpretation and reconstruction (see Fox, 1987a) than would a response like "No, I don't want to start with integrals." The fact that it appears to be taken as a kind of rejection in this case is very much a joint effort on the part of both participants.

Examples of some approach negotiations follow (*Zukowski* and *Washington* refer to calculus textbooks):

(7)
S: Mkay, so let's see now. (0.7) Ahm, maybe the best approach would be to- they like to give examples. And solutions and since I'//m so rough on this, maybe we should do something that elementary> Does that sound good?Or
T: Mhm
T: Sure.
S: Or, or if *you* have another idea, that's ah fine by me too, I'm a novice at this, so
   (1.2)
T: Ahm, (0.5) whatever you think you need help with

(8)
T: Where // shall we start?
S: ((laugh))
   (1.6)
T: Zukowski?
S: Uh (2.4) I guess. I- you probably don't know anything about this one right>Washington.
T: I could (0.6) take a look at it if you feel comfortable with that stuff. (0.9) Because (1.5) it's all about the same I would imagine.
   (1.2)

T: (CLEARS THROAT) So it does fractions (1.3) but you're mostly interested in starting in with integrals?
   (TURNS PAGES: 2.0)

T: or do you want to review some // other stuff first?
S: um (1.5) yeah, let's I think we could probably do a fairly quick review ah on in differentiation, (0.3) um
T: Okay

Once the tutor and student have negotiated an initial sense of how they are going to work together, they begin a new sequence of the dialogue, in which they initiate the tutoring itself. They use the tentatively formulated "plan" as a resource for negotiating who will initiate this new sequence: If the approach to the session has served to locate the source of trouble in an old exam or a set of already-worked homework problems, or exercises in a book—something that can be seen as coming in with the student—then it is appropriate for the student to initiate the new sequence:

(9)
S: Mkay, so let's see now. (0.7) Ahm, maybe the best approach would be to- they like to give examples. And solutions and since I'//m so rough on this, maybe we should do something that elementary> Does that sound good?Or
T: Mhm
T: Sure.
S: Or, or if *you* have an*o*ther idea, that's ah fine by me too, I'm a novice at this, so
    (1.2)
T: Ahm, (0.5) whatever you think you need help with
.

.
S: Okay. So let's see, uh, the id*e*ntities they said we had to uh (0.4) we had to memorize. // So I memorized some identities.
T: Yeah.
S: And those are the basic ones that uh (0.6) that he gave us. Okay. (0.4) And uh (LOOKS IN BOOK: 2.6) let's see (3.4) huh. (1.7) Okay. (1.0) Uhm, I don't know, what this is really asking, so I'm just going to (0.3) uh scribble on some paper, and see what happens, okay?
T: Mkay.
S: It's saying if s*e*cant of theta, (0.6) okay> now, secant of theta (1.7) equals three.

(10)
S: So we could just do the- like the pr*o*blems I've done most- I've done all my homework ((laugh)) I've been // good

T: ((laugh)) That's good.
S: And I've been doing *e*xtra problems. And- I brought the ones that I've been having a hard time figur//ing out.
T: Great.
    (1.1)
T: Great.
S: So (0.7) let's see (0.6) let's just start with chapter fifteen.

S: Number seven is the one I was going to do.

In the one case where the approach has been formulated not as working through problems that have come in with the student but as a review and then instruction on a topic new to the student, the tutor initiates the new sequence:

(11)
T: So polynomials // are okay.
S: I don't quite understand all that.
    (0.9)
T: Mkay. (3.2) Okay, so (1.1) chain rule?
    (1.5)
T: ring a bell?
S: Yeah, yeah, // chain rule rings a bell.
T: Okay.

The initial "plan" is thus used as a starting point for allocating responsibilities and dialogue roles. As a corollary to this fact, it is clear that not only are students active participants in the tutoring process once a tutoring sequence has been started; but they may also have some responsibility for selecting the problems to be jointly worked and hence for starting a tutoring sequence. As is seen in chapter 7, the student may also become actively involved in determining how to work the problem and for establishing what counts as an acceptable solution.

The last important aspect of the tutoring process that is negotiated in this opening segment is the level of tutor intervention, or, rather, tutor assistance. In some cases this level is overtly negotiated, as in example (12), whereas in others it is implicit in the emerging interaction, as in (13):

(12)
S: It's saying if *se*cant of theta, (0.6) okay> now, secant of theta (1.7) equals three. And theta is a fo*u*rth quadrant angle, find tangent of theta.
⇒   Okay, I'll just kind of mess around and you can tell me when I'm really off the track.

T: Mkay.
S: Did you want to do that?
T: Yeah.
S: Okay.

(13)
S: Number seven is the one I was going // to do.
T: (READING) How close must two electrons be if the electric force between them is equal to the weight, (0.7) of either at the earth's surface.
   (0.3)
T: Okay, so what did you:
   (0.5)
S: So this is what- we're allowed to have our little um, (0.4) sheet of formulas // so I don't have to memorize that, which is good. So this is the formula that I picked out.
T: Okay.
   (1.6)
S: It's uhm (1.3) What they want is the force.Right?
T: Right.

In (13), although the student never says to the tutor "be really interactive with me," and the tutor never says to the student "I'll help out when I see that you're having difficulty," it is clear from their utterances, including their timing and placement, that this is their emerging strategy: For example, after the tutor reads the problem statement aloud, it seems that she stops to let the student describe how she had tried to work the problem; there is no immediate response from the student, so the tutor prompts her to talk about how she had approached the problem. The student then starts in with this description. It is thus clear from their interactions that (a) the tutor is not going to let the student flounder for very long, and (b) the student is not going to brush off the tutor's attempts to provide assistance.

Needless to say, such decisions regarding level of tutor assistance are open to interpretation, reinterpretation, and renegotiation at any point in the process. Even in the one session in which the student explicitly asks to be allowed to flounder, for example, there are passages of high tutor assistance, as the following segment illustrates:

(14)
S: Okay. (1.6) Let me think for a second. (2.8) What if I factored through- if I times everything by cosine of theta, can I do that. (1.9) Like so?

T: .hh
    [
S: In order to factor (it) out?
T: You c*oul*d, but- there's something better to do right now. You can // add a common denominator. One over cosine minus
S: Mhm
S: mhm
T: sine times sine, (0.6) over cosine.

OPENING OF SEGMENTS

It is a major hypothesis of this study that the enterprise of tutoring functions chiefly as an arena for the contextualization of abstract formal concepts and notations for the student. Scientific and mathematical problem statements are highly dense sets of *instructions* masked as object descriptions, and instructions of all sorts are always underspecified with regard to the real course of actions they are meant to encompass:

> Social studies of the production and use of instructions have concentrated on the irremediable incompleteness of instructions (Garfinkel 1967, ch. 1), and the nature of the work required in order to "carry them out." The problem of the instruction-follower is viewed as one of turning essentially partial descriptions of objects and actions into concrete practical activities with predictable outcomes. . . . (Suchman, 1987, p. 101)

And scientific and mathematical problem statements are yet one step further removed from a description of a series of concrete actions than are typical instructions; they are usually not written in the form of imperatives (as in "remove tape before sealing") and do not specify a sequence of actions to be carried out. Rather, they tend to describe a collection of objects, either physical or mathematical, and the relationships among those objects. For example, a typical problem statement like "What is the speed of a 350 ev electron" is apparently a question about an object, providing no indication of action to be taken on the part of the problem solver. That the answer to the question can only be determined by working through some series of calculations—in more general terms, that the descriptions in the problem statement can in some sense be elaborated into a set of instructions—is a fact learned by the student through a long process of classroom socialization.[2] Moreover, the objects described in such tasks tend to be many levels of abstraction away from the concrete objects that the student may be accustomed to manipulating; they are often either entities that are not part of our conscious experience (e.g., electrons) or they are pure symbols (e.g., x).

Learning to see abstract problem statements as resources for situated activity is thus a difficult endeavor for most people, and that is why tutoring can be so valuable—it is designed to make the mechanisms of this process visible to the student.[3]

Consider the following physics problem statement:

What is the speed of a 350 ev electron?

---

[2]In everyday mathematical problem solving (e.g., in grocery store arithmetic; see Lave, 1988) the problem statement is often transformed directly into a description of the actions to be carried out, as in the now infamous "cottage cheese" example (see also chapter 7):

> Another problem posed to new members of Weight Watchers in their kitchens provides a further illustration. As in the exercise concerning peanut butter sandwiches, the dieters were asked to prepare their lunch to meet specifications laid out by the observer. In this case they were to fix a serving of cottage cheese, supposing that the amount allotted for the meal was three-quarters of the two-thirds cup the program allowed. The problem solver in this example began the task muttering that he had taken a calculus course in college. . . . Then after a pause he suddenly announced that he had "got it!" From then on he appeared certain he was correct, even before carrying out the procedure. He filled a measuring cup two-thirds full of cottage cheese, dumped it out on a cutting board, patted it into a circle, marked a cross on it, scooped away one quadrant, and served the rest. (p. 165)

Lave commented "Thus, take three quarters of two-thirds of a cup of cottage cheese was not just the problem statement but also the solution to the problem and the procedure for solving it" (p. 165).

[3]It is worth noting in passing here that one way in which the contextualization of abstract objects can be achieved is by turning them into active, sentient beings, with properties familiar to the student from experience with other sentient beings. Consider the following passage, for example, in which electrons are treated as having a particular gender, and feelings:

(1)
[they are discussing potential energy]
    T: But you make this vector r *so* large,
⇒ S: He can't
           (0.8)
    T: that r's infinity and this goes to zero. // So in other words if you put th*is* little q, (0.5) way out at infinity
    S: Right
    S: Aha
⇒ T: then the potential energy that he would have, due to th*is* charge, w//ould be nothing.
    S: Aha
    S: Oka//y
⇒ T: Because he doesn't even care if he's there, he's so far away.

The problem appears to ask a simple question; but, in fact, for most students the statement calls for an unpacking of each phrase, and their relationships to one another, into a series of equations. The student must be able to unpack "speed," for example, into some set of symbols that can be related to "electron volts" and must be able to formulate "electron volts" in some way that is relevant to the question of "speed." The student must be able to find the appropriate equation, find values for constants within that equation, know how to manipulate the symbols according to the rules of algebra, be able to calculate whatever numerical values arise, and so on. None of these actions are referred to explicitly in the statement; tutoring provides a collaborative environment in which students can develop the relevant practices for learning to use the problem statement as a constraining resource for some particular, situated, course of action, so that in the future they can perform this activity by themselves (e.g., in taking a test). It is therefore not surprising to find that much of the work the tutor and student do at the beginnings of sequences consists of contextualizing the exact phrases of the problem; it is, after all, at the beginning of sequences that the student must come to understand precisely what a problem "is asking" (as they often describe it).

The beginning steps of contextualization address the specific actions that are to "grow" out of a problem statement. One of these steps is understanding the deep conceptual organization of the problem: How do the superficial elements of the problem relate to this deep level? For example, after reading the problem statement just given, the tutor and student explore what speed is, and how it is different from velocity; they discuss the kind of units that are represented by *ev* and how those units are related to other, more familiar, units of energy.

As this last point indicates, another step is determining how the concepts of the problem can be related to each other and to other concepts the student knows. After the tutor leads the student to see that "electron volts" are units of energy, she asks the student how to relate energy to speed. And in contextualizing the concept of "electron volts," the tutor compares this very small unit of energy with the more familiar energy unit *joule*. Through these questions, the tutor guides the student to understand the fundamental relations of the problem, and thus how to set the initial equation up.

In working through this same problem, the tutor helps the student project what a reasonable answer would look like, long before they have come to the stage of calculating the final answer:

(15)
T: So you should make sure when you get your *a*nswer, that you d*o*n't get (0.6) 350 electron volts isn't 9 million joules.=
S: =Right. // It should be smaller.
T: You know
T: It should be a very small number of joules.

Another step of this initial stage of contextualizing, then, is working from the problem statement to a general format for a possible answer. This process helps the student know what he or she is working toward, and by implication, when to initiate major correction strategies.

Especially in physics, but in other scientific domains as well, technical terms that might appear in a problem statement also exist in English as nontechnical terms: *force* and *weight* are good examples from the domain of physics. As a result of this fact, the initial stages of contextualization are often taken up with clarifying the technical use of the relevant terms:

(16)
S: So I tried to look at the w*e*ight,
    (0.9)
T: And all's they give you is the m*a*ss.
.
.
T: You wrote down m*a*ss.
    (0.3)
T: Yeah, what's the difference between weight and mass.

As another step, the tutors seemed to be very concerned with explaining if and how a problem was similar to, or different from, previous problems that the pair had worked during the sessions. This concern is understandable: Problem statements that appear to differ in terms of surface elements may in fact represent similar conceptual structures and hence call for similar equations and solution strategies. Students must thus be helped to see what makes problems similar in a deep way, and this requires a good deal of contextualizing during the initial exploration of the problem statement:

(17)
T: Okay so let's see. [reading] Given three point charges located on the x axis, (2.1) what is the force on q two, due to the presence of the other two charges. And this is just the s*a*me thing.
    (1.0)
S: Mhm
    (0.8)
T: (READING) Given the magnitude of the // (natural force)
S: But (now) I don't have to do- a//dd them vector//ially, yeah.
T: You have no angle.
T: Right, everything's on the line.

In this passage, the tutor compares the current problem statement with the problem they have just worked: "And this is just the same thing." Interestingly enough, it is the student who steps in with the understanding that the problems, although similar, are not identical: In the previous problem, they had to add the forces vectorially, but because in this problem the charges are all on the x axis, there is no need for this procedure. The problems are thus the same in some ways and different in others. This passage provides a compelling illustration that establishing "sameness" of problem statements is part of the initial contextualization that tutors and students proceed through, and that what "same" means in any particular case must itself be contextualized by further interaction between the tutor and student.

The following passage clearly illustrates the general practices a tutor can use to unveil the actions that lie behind the question–answer structure:

(18)
T: So let's do one of these. (1.3) We'll start with the simple ones. Five ex minus twelve over ex times ex minus four. First check the degrees> the numerator is smaller than the denominator so that's okay> factor (0.2) they already did that // for us.
S: Right.
T: And then when you break it up (0.3) th*is* stuff (1.1) is going to have to be equal t- for *ex* we have *o*ne power, so ey over ex

T: We want these two things to be equal (0.7) so- the f*i*rst thing we could do is just make them look more alike (0.7) get the common denominator. (0.3) This is a step that you will skip after a few times, but this is (0.2) where it comes from.

In this example the tutor takes an entirely abstract, formal, statement ($5x - 12/x*x - 4$) and contextualizes it into specific, ordered, actions.

Another example follows:

(19)
T: When you l*oo*k at this, you know you got this 300 electron volts, and you go- and you always go, oh my God what (0.4) what is an electron vo//lt
S: And what *is* an electron volt
T: and then if you can *a*ny way fool around with the units, to figure out well what is it I'm talking about you know, and you go well it's q times v and you look over at your equation and you go okay well that's (0.2) w*or*k and energy are the same thing.
S: Aha

T: And then you go, *energy*, okay, well wh*at* energy do- equation do I kn*ow* that has velocity in it.

Notice in both passages, that the tutors formulate their ordered steps to straddle the line between *types* of actions and *particular* actions; that is, they are showing the students how in general to proceed with this class of problem while at the same time showing them how to proceed on this particular problem. This kind of approach is common in our tutoring sessions, presumably because (a) tutors want students to be able to solve any problem, regardless of its superficial features, if it belongs to a class of problems they have worked through, and (b) tutoring is designed in part to make the relationship between surface features and deep conceptual organization transparent to the student.

Later on in the session excerpted in (18), the student is given the opportunity to be the main problem solver. Notice in the following passage that the tutor now uses the step-by-step procedure she introduced earlier as a scaffolding structure within which the student can situate his actions:

(20)
T: Maybe nineteen. That- that has a good factoring technique we can look at.
    (1.2)
S: (SNIFFS)
    (7.1)
T: Of course you probably have to use another trig substitution.
    (13.1)
⇒ T: Okay, so the first thing's // to
S: (Star) ready.
    (0.6)
T: Good.
    (2.7)
⇒ T: Second thing> is> (1.0) to>
    (1.7)
T: fa//ctor>
S: factor the denominator.
T: Right.

Part of this process of proceduralizing the question–answer structure, which is especially prominent at the beginning of tutoring segments, is the classification of a problem statement within a typology of problem statements (where the full typology may or may not be made explicit). In this process, the student learns to see which surface features of a problem statement, for the purposes of a particular kind of problem solving, count as distinguishing features and which count as insignificant "noise."[4] An example follows:

(21)
T: (READING) Given three point charges, fixed at thr*ee* corners of a square. Find the electric field and tendency vector *ee*, at the corner p*ee*, with no charge, this is exactly like the problem we just did
⇒ S: Right.
T: And show the direction with a *ca*refully drawn *a*rrow>

In the next example, the tutor has already told the student that determining whether the function to be integrated is rational gives one a handle on how to approach the problem; so one of the first things she comments on in setting up the problem to be worked is that it is a rational function:

(22)
T: Let's look at twenty (0.5) um five looks like a good one. This (won't) get us to do all kinds of stuff at the same time. Okay, it's a rational function and before we get into it let's check the substitution.

It is fairly apparent from these examples that the joint contextualization of problem statements (and formulas, symbols for constants, etc.) represents *collaborative cognition* (Fox & Karen, 1988: Lave, 1988; Rogoff, 1990), by which is meant the making public of what will become for the student an internal cognitive process. In situations of explicit collaborative cognition,[5] the tutor, with help from the student, makes overt the social practices that lie behind problem solving. In this way, the tutor and student together produce the processes that the student will eventually be called on to do by him or herself (e.g., on an exam). This phenomenon of collaborative cognition has not, to the best of my knowledge, been documented before for any adult interaction and has not been addressed in the tutoring literature (something like it has been discussed in the language development literature; see Ochs, 1988; Vygotsky, 1978; and for other adult–child interactions, see Rogoff & Lave, 1984).

In recent work, Dellarosa (1986), and Dellarosa, Kintsch, Reusser, and Weimer (1988) have shown that young children's difficulty with arithmetic word problems lies in their lack of linguistic sophistication, their inability to go from

---

[4]Presumably much the same process occurs when a child learns which phonetic features count as making a difference (constituting different phonemes) and which are irrelevant or predictable from context (constituting the same phoneme).

[5]It could be argued, and in fact it has been argued (see, e.g., Ochs, 1988; Suchman, 1987; Vygotsky, 1978), that all cognition is internalized collaborative cognition, and hence even as internalized processes retains elements of social collaboration.

the linguistic description (the word problem itself) to the desired conceptual representation. The present study both supports this finding and challenges it. It supports the finding in that the students in our study clearly had difficulties with what one might loosely describe as "understanding what the question was really asking for"; that is, they could not easily go from reading the problem to setting up the answer, even if they in fact knew how to solve such a problem. This indicates that Dellarosa and Kintsch are exactly right: What causes difficulty is the mapping, if you will, between the text and the conceptualization, and between the conceptualization and a course of action.

However, given that our students are college-level, taking difficult courses, we cannot assume that the trouble arises from a lack of linguistic sophistication; by all measures, these students are fully developed linguistically. Perhaps college students find word problems problematic precisely because the relationships between the problem statements of word problems and the course of action that they are meant to project are not just underspecified, as with common instructions—they are, in fact, for the most part *un*specified.

Our tutors seemed to recognize this fact about problem statements, and each treated situating a problem as a trial-and-error process; that is, they treated the surface features of the problem statement as not in any sense uniquely specifying the "correct" initial course of action but as a suggestion of a range of possible courses, some that may prove to be inadequate only after the problem solver has explored them. Thus, initiating work on a problem can be seen as an iterative process, where one path can turn out to be a dead end, thereby requiring the problem solver to go back to start again on another path. And the tutors also seemed to recognize that this aspect of problem solving is not a difficulty to be gotten over with increased expertise; rather, it inheres in the nature of the relationship between problem statement and situated action:

(23)
T: And *o*ne thing- when you start problems?
S: Mhm
T: It's okay to just write down four or five equations, y//ou know, and just kind of look at them.
S: Okay.
S: Yeah.
T: It's a- even if you- because you're trying to figure out what you're doing.
S: Yeah, ((lau//ghter)) That's true.
T: ((laugh)) It's okay to kind of bungle along, at first, you know.

(24)
T: You must have a way to solve systems of equations.
    (1.1)

T: Some favorite way
   (1.2)
T: I use the (0.2) punt method and just
S: Punt method ((laugh))
T: start working. This one's going to be easy. So I find the easiest one.

In many cases the tutor led the student down a garden path by starting on an inadequate hunch, illustrating the point that one cannot always determine in advance which path will yield fruit. The tutors often made their own "mistakes" the source of some reconstructive repair work that the student could benefit from:

(25)
T: You know, this is *a*wfully funny, isn't it?
S: ((laugh))
T: Oh, you're goin- you're going to l*o*ve this when we get to the end, we should've noticed this right away
.
.
T: You're left with the stuff you started with. And we should've actually noticed this sooner (1.0) That *is* the decomposition.
T: This is something that you hopefully avoid right away.
   (1.0)
T: What should we have done. (2.7) Now, i- if you do the partial fraction decomposition and you find out that- it *is* already decomposed, then that means that- you have some other technique to do it, something that we've already learned how to do. So, what's the only other thing we know how to do.

(26)
T: (READING) A *o*ne *o*hm w*i*re. Okay, what's *o*hms
   (1.4)
S: It's resistance
T: Okay. Is dr*a*wn out to thr*ee* times its original length. What is the resistance now.
   (2.1)
T: I think that's a hard question.
S: Yeah.
T: I would say it was the same.
   (0.7)

S: I would too, then I looked it up in the back of the book and it said nine times
       (1.0)
T: Oh really?
.
.
.
T: Now, you know now that I think about it. Okay let's say you were an electron, okay?
S: Aha
T: And you had to fit through a p*i*pe
S: Aha
T: Okay? Would it be *ea*sier for you to go through th*i*(h)s pipe, (0.9) or th*is* pipe.

(27)
T: (LOOKING AT HER OWN WORK) Oh-oh, there's a problem here.
       (2.8)
T: That's not right.
       (2.5)
T: One over (1.8) hmm (1.0) over
.
.
T: I know what I'm gonna do. I'm gonna come back up here
S: Oh huh
T: and uh simplify all this and then take the inverse of that. That would be easier.

Given the opaque nature of problem statements, such garden paths are an "irremediable" (Garfinkel, 1967) part of real-time problem solving. It is thus impossible, as the tutors recognize, to hide this aspect of problem solving; and it is also unwise to take on these "mistakes" as indications of failure or lack of expertise. An integral part of problem solving, these garden paths are thus treated as valuable learning experiences for both the tutor and the student.

This chapter has explored the openings of sessions and the beginnings of segments within sessions. The next chapter examines the nature of correction once the session is underway.

5 Correction in Tutoring

GOALS OF THE CHAPTER[1]

As the tutoring sessions move from the opening into the serious business of problem solving or instruction, the need for correction and intervention, by both the tutor and student, becomes a central focus for the tutoring dialogue. This chapter describes the mechanisms whereby correction and intervention occur in the tutoring sessions.

I have chosen to focus on correction and intervention in this chapter not only because of their central role in tutoring dialogues, but also because of the importance they have attained in the ITS literature. Some systems, for example the systems developed (following the ACT* theory) by Anderson and others (e.g., Anderson, Boyle, Farrell, & Reiser, 1984), already assume a model of the interactional aspects of correction based on intuitions about what is psychologically effective for students. The research presented here provides empirical evidence for how tutors and students actually do carry out such necessary tasks as correction.

ORGANIZATION OF TUTORING INTERACTION: CORRECTION AND TUTOR ASSISTANCE

The main focus of this chapter is the process of a subcomponent of repair, namely correction, including the process of tutor intervention, or, as I prefer to

---

[1] An earlier version of this chapter appeared in Goodyear 1991. I would like to thank Ablex for permission to use it here.

think of it, tutor assistance. *Tutor assistance* is the superordinate term here for all forms of help from the tutor, including help that is oriented to steering the student in the right direction. *Repair* is subordinate to tutor assistance and is the general term for fixing any prior problem in communication, and does not imply that the problem is a result of error; *correction* is a subtype of repair, covering repairs that arise out of errors.

We have found that the best way to understand correction (including tutor assistance) is to examine the overall structure of the tutoring sessions; in other words, correction is best studied within the sequential organization of the interactions themselves. The full structural details of each interaction are beyond the scope of this chapter; here I provide only enough description to motivate the analysis of correction.

As we saw in chapters 3 and 4, the sessions divide themselves into two groups: The first group is characterized by the student working through problems in a textbook (in one case the student brought in a set of problems from the text that she had been having difficulties solving); the second group is characterized by something more of a lecture format, wherein the tutor talks about domain concepts and demonstrates how to solve problems.

As we saw in chapter 3, when students work through a problem with a tutor, they very often verbalize what they are doing, step by step. For example, in the following excerpt, the student doesn't just write down the steps as he works through them—he talks about what he is doing as he does it:

(1)
S: So:, ah:m (1.0) this, okay, secant, of theta, I know eqs- equals three:. Now, so but- my equation here is (0.4) secant squared theta. So what I want to say is .hh the square root, (0.6) of secant, (0.5) squared theta, .hh would equal three:. Right?

Why do students produce this on-going commentary? They do it to display to the tutor how they have understood the problem and how they understand what they are currently doing to solve the problem. So every student utterance of this sort is a display of understanding. As with any understanding, the student's understanding of a particular step in a problem can match or not match the interlocutor's understanding, so that any such display of understanding calls for a confirmation (or disconfirmation) from the tutor, by which the tutor displays that the student's displayed understanding matches her own (see Labov & Fanshel, 1977, for a related point). A confirmation agrees with the student's understanding; a disconfirmation disagrees with it.

From past research (Pomerantz, 1975) we know that agreements come very quickly after the utterance they agree with, whereas disagreements are somewhat delayed. In these tutoring sessions we have found that when the tutor agrees with the student's displayed understanding her signal of confirmation comes quickly after the student's turn:

(2)
S: Mkay. .hh And I know it's negative, just to follow your thought process, because I know that the sine is positive.
T: Mhm

(3)
S: And this (draw it out). (0.3) And the double bond goes away
T: Right

Whereas if the tutor disagrees with the student's understanding, the delivery of the disagreement is somewhat delayed and in some cases what might be described as "hesitant":

(4)
S: And it's going to change when I put this in- there, right?
    (1.7)
T: I don't think so.

(5)
S: So that triple bond is like ess pee three?
    (1.1)
T: Ah:: no:, that- a triple would be an ess pee.

Thus, tutor correction or indication of a problem with the student's understanding is delayed with regard to the relevant student utterance.

Furthermore, it follows from this and past work on disagreements (Pomerantz, 1975) that when a student has produced a display of his or her understanding and there is no immediate response from the tutor, the student can anticipate that the tutor is going to disagree with his or her understanding; students in this situation will very often re-phrase their statement as a question, thereby *inviting* correction from the tutor:

(6)
S: Okay, just for review for my sake, .hh a cosecant is .hh uh:m, one over the tangent.
    (1.3)
S: Am I correct?
T: N:o.

(7)
S: And it's going to change when I put this in- there, right?
    (1.7)
T: I don't think so.
    [
S: Does the capacitance change?
    (0.5)
T: I think the charge changes.

Students can also use this "predisagreement" silence to try by themselves to correct whatever may be wrong with their understanding:

(8)
S: I use this one for that one, 'cause I don't think I had a dielectric
⇒   (0.6)
⇒ S: wait, no, I do:
        (0.9)
T: ts 'cause you have paraffi//n
S: Paraffin.
T: That's the whole point about the paraffin.

(9)
S: ey: is minus one, and bee is zero.
⇒   (1.5)
⇒ S: No, bee is one.

If the student's understanding does not match the tutor's understanding, then, there are three possible outcomes, listed here:

1. the student can correct him/herself
2. the student can invite correction
3. the tutor can initiate correction

A note is in order at this juncture regarding my use of the word "match." I do not mean for this to imply that the tutor has an abstract understanding that she compares with the student's to find discrepancies; if this were the case, and the tutor were sophisticated enough and the student novice enough, then everything the student said would have to be corrected, inasmuch as a sophisticated understanding of some domain is apt to be radically different from a novice's. My own understanding of a phoneme, for example, is not what I would bring to bear in tutoring beginning linguistics students. It must be, then, that the tutor takes into account the task at hand (e.g., solving a textbook problem), and the kind of understanding that one could expect from a novice, to determine what counts as "wrong" for the purposes of this particular interaction. The importance of this point cannot be overstated; it re-appears, in a somewhat different context, in chapter 7.

The student and tutor go on in this way, displaying and repairing their understandings until the student gets stuck.[2] Here again there are several alternative responses that either participant could make, assuming that the student is displaying overt signs of "being stuck," and not, for example, of "thinking."

The situation of a student being stuck and showing being stuck creates a potential conflict. In our everyday interaction, if we see someone having difficulty in some way, it is preferred for us to offer help before that help is requested (Pomerantz, 1975). I do not mean that it is personally or psychologically preferred for us to offer help in this situation; indeed, we may be in a hurry, or not like the person, or have something else we would rather be doing. Rather, it is preferred socially and structurally, so that if we are not going to offer help we must provide an excuse for not doing so, or pretend we did not see the trouble, and so on. Preference organizations of this sort are independent of the momentary preferences of individual participants.

It is possible that in tutoring interactions the preferred response to the student's difficulty would be for the tutor to in some way offer help. But, as we have seen, there is a conflicting preference organization that indicates that participants should be allowed to repair their own trouble (Schegloff, Jefferson, & Sacks, 1977).

The tutors in this study display an orientation to both kinds of preferences (for a similar finding with regard to compliment responses, see Pomerantz, 1978).

Tutors in our study did provide assistance, but they did so in such a way as to give the student the opportunity to unstick themselves, both before the assistance is provided and while the assistance is emerging.

(10)

S: and they want to know what the tangent is. So, I have one over cosine of theta equals three. (0.8) And I have the sine of theta over cosine of theta (1.0) hmm:. (0.8) .hh Okay, so I guess I somehow have to: (0.8) tangent

---

[2]In some cases the student avoids getting stuck by indicating before the difficult step that this next step was exactly the reason he or she needed help with this problem:

S: And kay is the: constant that, I know that (0.5) and um (1.2) eff, that's what I had a problem with, was eff

This strategy only works if the student has worked the problem before coming to see the tutor.

of theta is going to be: (0.4) sine of theta over cosine of theta. (2.0) One over cosine of theta,//so (0.3) three.
T: Mkay. Now,
S: Okay
T: ts looking up here, ju//st at what=
S: Aha
T: =they've done, (0.4) .hh cause I can tell, we're headed in the wrong direction.
S: Ye:ah, they used to con-they use // one of the pythagoreans.
T: One plus (0.6) tangent squared e-quals the secant squared
                                              [
S: secant squared.

The student in this case is going around in circles—repeating that tangent equal sine over cosine—without finding a new way to look at the problem. The tutor intervenes, but not without giving the student a fair opportunity to figure out the answer for himself. Furthermore, the assistance is produced in such a way that the student can collaborate in the redirection, as he in fact does with his last lines. Here again, correction/assistance is momentarily withheld to give the student a chance to fix the problem himself. The withholding time is not long, however, and a student who wishes to be given a longer opportunity to work the problem out for him or herself must specifically request such an "extension":

(11)
S: Now, .hh let's see, when we said tangent of theta was less than zero .hh u::hm the tangent was
        (0.7)
S: give me a second. The tangent was sine over cosine.
T: Mhm

Correction/assistance of this sort thus is slightly delayed but is still offered without being "overtly" requested.

Tutors regularly provide assistance for the student if the student has produced one step in a chain of reasoning but apparently does not see the inference(s) that should be drawn for the next step:

(12)
S: Ey plus cee equals zero.
        (0.3)
⇒T: Right, so that tells // you

```
     S:  Ey equals cee.
              (0.4)
     T:  Minus cee.

     (13)
     S:  So it's got to be: in our fourth quadrant.
     T:  Right.
              (1.4)
     S:  Aha. (0.8) Aha, o//kay.
⇒ T:  Which means?
     S:  Then I have to come back down here and I'm- you're asking me to choose
         a sign, right//t?
     T:  Right
     S:  .hh Okay.
     T:  For your cosine.
```

The conflict between the tutor providing help and the student working through the trouble him or herself is overtly displayed in the following passage. The fragment starts with the student trying to determine the quadrant for the tangent given in the problem; he first gets into trouble by giving the wrong formula for cotangent. The tutor provides help and then the two appear to play a very tame kind of tug of war to see who is going to do the next steps of the reasoning.

They subtly try to wrestle a few turns from one another, culminating in the student saying "let me see if *I* can figure that out."

```
     (14)
     S:  I have to place it in a quadrant, is what you're telling me, right?
     T:  Mhm
     .
     .
     .
     S:  I would say: (.) uh, a cotangent, in terms of ex wai, (0.3) is let's see, one
         over the uh- cotangent is one over the uhm, (0.7) hold on a sec,
         (LAUGH) uh: cotangent is one over the sine.
              (0.4)
     T:  No:
     .
     .
     T:  Cotangent is one over the tangent.
```

S: Now, if I'm thinking in terms of ex and wai, though, (0.8) fo:r (0.8) the sake of the quadra//nts>
T: It would be cosine over sine.
 (0.8)
S: Right=
T: =Which is ex over wai.
S: Okay
.
.
T: And the cosecant//is-
S: Co- cosecant, it's that's the one over the sin//e right
T: One over sine.
.
.
T: Which means that your ex value is positive.
 (0.2)
S: Right.
T: Which puts you in:
 (0.5)
S: 'kay, .hh let's see if I can figure that out.
T: Okay.

Often, especially in more conceptual domains such as physics, tutor assistance is provided in the form of a question whose answer will serve as a resource for getting the student unstuck. This strategy has two parts: The first part requires that the student be able to answer the question, and the second requires that the student see how that answer is a resource for continuing the problem. Because both of these processes may end up involving correction, and because correction is dispreferred (Pomerantz, 1975), this strategy is undertaken very cautiously and with a heavy degree of support from the tutor:

(15)
S: eff, that's what I had a problem with, was eff, they said (3.1) if (0.5) if the electric force between them is equal to the weight>
 (0.4)
T: ts Okay.
S: So: I tried to look at the wei:ght,
 (0.9)

T: And all's they give you is the ma:ss.
    (0.2)
S: and it- yeah: (0.7) Oh, that's what it was, it was the mass.
    (0.8)
T: Yeah:.
    (1.0)
S: Oh:, I s//ee, I want weight.
T: You wrote down mass.
    (0.3)
⇒ T: Yeah, what's the difference between weight and mass.
S: I used to know this let's see
    (1.7)
T: I think, (0.8) I think what it is (0.8) is that, (0.5) n- what is uh, when you do (0.6) uh:, gravity problems.=
S: =Right.//It's that-
⇒ T: What do you always do?
    (1.2)
S: You have to multiply it by the-=
T: =by gee.=
S: =Gee. Right.

.
.
.

S: So I need to multiply this time (0.8) gee.
T: Right.

There are two tutor assistance questions in this passage, marked by arrows in the margin. The second question is of course meant as a resource for answering the first question. Notice that the tutor gives the student a fairly long space in which to answer the question, and when the student shows that she is not able to answer, the tutor does not directly provide the answer. Rather, she asks another question, whose answer and import will enable the student to answer the first question. The student and tutor work together to produce the answer to the second question, at which point it is relevant for the student to indicate that she sees the import of the question for the problem they are trying to solve.

The student eventually sees how the answer is a resource and goes on to the next step of the problem. The tutor provides a safety net around the student, so that if she shows signs of not being able to answer the question, the tutor offers a resource for answering. If the student shows signs of not seeing the import of a question for the problem at hand, then the tutor steers the student toward seeing the connection. All of this is kept in balance with not correcting or redirecting the student before she has had the opportunity to do those things herself.

Tutors also ask questions before the student has gotten stuck, often to help frame the problem, and the solution.³ In the following passage, for example, the tutor checks to see if the student understands the kinds of units—and therefore the appropriate formulae—the problem involves:

(16)
T: (STARTS READING THE PROBLEM) A one ohm wire.
⇒ Okay, what's ohms
  (1.4)
⇒ S: It's resistance
T: Okay. Is drawn out to three times its original length. What is the resistance now.

The response to these questions is carefully monitored, as with all tutor questions, to make sure that the chances of the student producing an appropriate answer are maximized. If the student has difficulty answering the question, the tutor will provide clarification, hints, and so forth, as in the following example:

(17)
T: What is the speed of a three hundred and fifty ee vee electron.Okay.
T: So the main thing he:re, I mean, when you look at that, what is electron volts, what kind of a
S: It'//s ahm,
T: what are we talking about.
  (1.5)
S: Isn't it- the charge of an electron times?
  (0.9)
T: Right,//but what is,it-what is that. Is it-=
S: The voltage?
S: =It's s:maller
  (0.2)
S: I//t's

---

³This framing is reminiscent of Vygotsky's concept of *scaffolding* (see Vygotsky, 1978), by which the teacher in an apprenticeship situation structures a task such that it is always within the learner's abilities; as the learner's abilities increase, the teacher gradually removes the "props" which have made the task approachable, until finally the learner can perform the task without assistance. See also A. Collins, Brown, and Newman (1986).

T: No- okay, I'm n- I'm a I'm a I'm not asking a specific enough question. U:hm, (1.1) Uh (0.3) is this units of length?
   (0.9)
S: Oh, no it's uhm
   (1.9)
S: It's voltage, isn't it?
[the tutor goes on to redirect the student until she sees that the answer is energy]

In her study of human tutoring, Galdes (1990) found a slightly different pattern of tutor assistance in this regard, which raises important issues about context and tutoring strategies. It is worth exploring her findings.

First a discussion of the differences between her study and ours is in order: Galdes did not have a face-to-face arrangement; she put the tutor and student in separate rooms, allowing them to communicate by means of an audio system that could be turned on and off. The tutor also had visual access to what the student was doing (just the student's hands and work area), by means of a video hook-up. The student had visual access only to information that the tutor posted on a board. When the tutor and student were not directly engaged with one another, the audio channel was turned off, and had to be turned on again by either the tutor or student to reinitiate conversation.

Galdes found that when the tutor and student were not engaged in a conversation (i.e., when the audio channel was off) the conflict for the tutor between offering assistance and letting the student ask for help was somewhat less pronounced:

> *Factor #14 - The likelihood of whether or not the student will call soon.* The tutor considered this factor when the category of the definite error was that the student appeared *generally confused* such as tapping his pen, continuously shuffling his lab results, or doing nothing for several seconds. For these instances, if it appeared that the student was *likely to call soon*, the tutor waited to interrupt. (Galdes, 1990, p. 245)

Although this finding is not terribly different from what we found in our sessions, it is noteworthy that Galdes' tutors often sat back and waited for the students to call to ask for help, whereas our tutors were more likely to step in with some form of assistance. The reason for this difference is not clear: It may be the result of the face-to-face nature of our interactions. The same underlying principles and struggles are at work here; what is different is the organization of the tutor–student interaction, which then shapes the strategies the participants adopt to negotiate the conflicts. In the face-to-face condition in our project, because the tutor and student were always conversationally engaged with one another, there was never a time during which the student was working alone without also interacting with the tutor. In Galdes' study, on the other hand, the student and tutor in this situation were not interactionally engaged with one another. In our project, the tutor had only to speak again to offer assistance; in Galdes' study, the tutor had to initiate a new conversation to offer assistance. The sense of interruption is thus much greater in the latter than in the former; it is little wonder that tutors hesitated to use such a powerful mode of interruption to a greater extent than they did in our study. The difference between the two studies illustrates the subtle ways in which the organization of interaction in a tutoring session (or in any interactional setting) affects the conversational patterns of the participants.

Returning now to our own study, for the sessions in which the tutor does most of the talking (both about general procedures and concepts and about solving problems for the student), the student can display his or her understanding by finishing the tutor's utterances, as in:

(18)
T: So this is really the integral of ex (0.8) the who- e- ex goes in the whole time=
 [
S: Right
T: =.hh plus the remainder one over ex.
 [
⇒ S: one over ex. Yeah

(19)
T: In order for this to become basic,
   (1.2)
⇒ S: It'll have to lose a=
T: =it'll have to lose a hydro//gen
⇒ S: hydrogen

In fact, the tutors often capitalize on collaborative completion as a means of finding out what the student understands by starting an utterance, with a slightly rising intonation, cuing the student to finish the utterance appropriately:

(20)
T: So I need to: do what.
   (2.0)
⇒ T: Multiply on the inside by
   (0.1)
S: half

⇒ T: one half to get rid of that. (0.8) And so on the outside I'm going to be:
    (1.3)
⇒ T: multiplying by>
    (1.7)
   S: a factor of two, yeah
                [
   T: two

This strategy, which uses statement syntax to elicit information from the student, neatly avoids the problem of correcting a wrong answer from the student —if the student provides an inappropriate completion for the utterance, the tutor can provide the "correct" answer as if she were merely finishing her own sentence.

(21)
T: Integral of ex gives us ex squared over two> One over ex gives
    (0.4)
S: One ex
    (0.5)
T: Natural log
    (0.4)
S: Absolute value of ex

The tutor can also incrementally add clues to the partial utterance if the student fails to finish the utterance at the first opportunity:

(22)
T: .hh Second thing> i:s
    (1.0)
T: to>
    (1.7)
T: fa//ctor>
S: factor the denominator.

In this passage the student is unable to complete the tutor's utterance until he hears the first syllable of *factor;* as soon as he is equipped with that clue, he is able to produce the appropriate completion. This strategy provides the tutor with a kind of metric for judging the student's understanding, inasmuch as each opportunity to complete the utterance that is passed is some indication of how well the student is keeping up.

In some rare cases, the tutor can even end up completing her own utterance entirely, if the student fails to provide a completion:

(23)
T: the main thing with the exponential function, (1.8) you have (0.2) ee to the ex (1.4) and its derivative is
(2.3)
T: ee to the ex

The utterance-completion strategy tends to be used mostly for working on a step of a problem (usually a problem being solved out loud by the tutor in front of the student). For more conceptual issues, or more tactical issues, the tutor will often ask a question, with the syntactic form of an interrogative:

(24)
T: Is there something we can put in here, to wipe this part out?=
S: =Well that one will go to zero, yeah.
T: Yeah (//)
S: The ex equals zero.

Here, as with the tutor questions discussed previously, the issue of correction is prominent. This format is overtly a question, which makes an answer from the student socially appropriate. Both parties work to avert an incorrect answer from the student. They manage this in an intricate way: The silence that grows after the tutor's question is carefully monitored by the tutor for signs that the student either will or will not be able to answer the question. The student participates by displaying signs of working on a calculation, for example, or by displaying signs of confusion or lack of comprehension. The tutor participates by looking for these signs and responding appropriately.

For example, if the tutor has asked a conceptual question and the student responds by "staring blankly" at the textbook, the tutor is likely to provide assistance before the silence has grown much beyond 2 seconds; on the other hand, if the tutor asks the student what the outcome of a particular calculation is, and the student responds by displaying signs of "working on it," the tutor may allow the student a fairly long silence, usually lasting until the student either answers the question or displays further signs of being stuck. Eye movement, facial expression, body posture, position of pencil, and nonlinguistic verbal cues (such as sighs, inbreaths, clicks) are all used and monitored during this particularly sensitive time.

There are three central outcomes of this interaction:

1. the student answers the question (usually correctly);
2. the student asks a question about some portion of the problem or question, the tutor answers and the student then answers the original question; or
3. the tutor provides assistance—in the form of clarification, hint, a more leading question—and the process starts again, most often until the student answers. In a few rare cases, the tutor provides the answer itself, but this occurs only after the student has passed several opportunities to answer.

The most striking outcome of this whole process is the low rate of incorrect student answers: I found only 12 incorrect answers produced after tutor questions, out of 97 possible answer slots. This low rate is seen as an achieved outcome of the processes described earlier, rather than a natural fact of, say, the student's IQ or knowledge of the subject matter.

Summary of Correction Strategies

We have seen that the tutors in our study withheld correction and even assistance until the student had an opportunity to initiate correction on his or her own. In some cases, this strategy is effective in getting the student to produce a correction, but in other cases the student is unable or unwilling to produce a correction; in these cases, the tutor takes on, in some instances with the overt collaboration of the student, the task of correction or assistance. How the trouble is handled depends on where in its sequence it is produced, in the following way.

There are four main positions in which the tutor engages in correction, or initiation of correction. In the first case, the student has produced a display of understanding that is in some way incorrect; the tutor withholds correction and in this space the student, anticipating disagreement from the tutor, invites correction. The tutor responds with a correction.

(25)
S: Because secant squared of theta is square root of
   (0.8)
⇒ S: Can I do it that way?
S: S- can I say three minus one?
   [
T: Mm::
T: No, you want to say three squared. Because the secant is three.

The second position is the case of a wrong answer produced after a tutor question. The tutor in this situation regularly initiates correction and the student attempts self-correction.

(26)
   T: Did the area change?
        (1.0)
   S: Wouldn't the area be the same?
        (0.9)
⇒ T: We only have the same amount of copper.
   S: Yeah.
   T: Well think of taking silly putty.
   T: Like a block of silly putty like this> and you pulled it out? What would ha//ppen?
   S: It's the same.
        (0.2)
   T: It would get long and skinny, though.

Notice that when the correction is not invited by the student, the tutor does not overtly correct the student; rather the tutor tries to redirect the student's thinking. The behavior of the tutor is thus clearly sensitive to the context of utterance of the problem.

The third position follows an utterance or set of utterances by the student that usually exhibit(s) being stuck. In this position, the tutor regularly initiates correction and allows the student the opportunity to accomplish the final correction:

(27)
   S: tangent of theta is going to be: (0.4) sine of theta over cosine of theta. (2.0) One over cosine of theta, // so, (0.3) three.
⇒ T: Mkay. Now,
   S: Okay
⇒ T: ts looking up here, ju//st at what=
   S: Aha
⇒ T: =they've done, (0.4) .hh Cause I can tell, we're headed in the wrong direction.
   S: Ye:ah, they used to con- they use one of the pythagoreans.

(28)   ((Student working problem: 10.8))
   T: .hh Where did this minus sign come from.
   S: .hh It's minus ex. This minus ex shouldn't be here.

The fourth position is the rarest, and involves the student producing an utterance—usually in conjunction with working a step of a problem—that completes a tutor prompt. In this case, the tutor initiates and accomplishes correction by simply producing the rest of her original utterance, with the correct piece of completing material serving as an embedded correction. These instances all involve low-level calculations that the student has produced while working a problem.

(29)
T: One over ex gives>
    (0.4)
S: One ex.
    (0.5)
T: Natural log.

We can see from these examples that where an error arises affects very much how it is handled by the tutor and the student. Thus, in order to understand the kinds of strategies employed in tutoring, we need models of conversational structure in tutoring.

CONCLUSION

In addition to the implications for modeling correction and tutor assistance, the work presented here brings a new emphasis to the role of the student in the learning process (although see Cumming & Self, 1991; Miyake, 1986; Oberem, 1987; Woolf, 1984, for suggestions in roughly the same direction). In particular, it should be clear from this work that, given the opportunity, learners play a critical role in the structure and substance of a tutoring session.

It is important here to stress a view of tutoring as an accomplishment achieved by both parties, a cooperative endeavor. Any outcome of the tutoring process (e.g., a low rate of student error, or an interaction that is more tutor-guided in some places and more student-guided in others) should be seen and analyzed as a product of the work of both parties.

6 Interaction as a Diagnostic Resource in Tutoring

TIMING AND DIAGNOSIS

As the tutoring sessions progress, and the tutor and student begin to form more elaborate models of one another's knowledge, the process of diagnosing student difficulties, in tandem with the correction process described in chapter 5, assumes a major role in the tutoring, and increases in detail and sensitivity as the sessions evolve.

A good deal of research in ITS has focused on the diagnostic process, but most of it has explored bugs in students' problem-solving behavior. According to much of the recent work in ICAI (see Anderson, Boyle, Corbett, & Lewis, 1987; Oberem, 1987; Sleeman & Brown, 1982; Woolf, 1984), the main job of a tutor is to discover and repair "buggy" procedures in a student's work. As we have seen, tutors appear to be oriented to a variety of tasks not centered on bugs, but nevertheless, tutors do spend a great deal of time guiding students to reconstruct misconceptions and faulty procedures.

In much of the ITS literature, tutors are said to derive their knowledge of student bugs by comparing the student's problem-solving procedures to their own (e.g., Anderson, 1988; Sleeman & Brown, 1982; VanLehn, 1988; Wilkins, Clancey, & Buchanan, 1988); in other words, an examination of the student's steps in solving a particular problem is the major source of diagnostic information. There issues from this assumption a second assumption, namely that a successful ITS should be able to carry out debugging procedures by examining the steps a student enters while solving a problem.

I would like to present here a somewhat different view of real-time diagnosis, and more generally, of assessment.

INTERACTION AS DIAGNOSTIC RESOURCE

That tutors base some of their "debugging" efforts on steps in solving a problem is uncontroversial. However, there is another source of diagnostic information used at least as much, if not more than, the steps in problem solving, namely, the tutoring interaction itself. Tutors use the timing of a student's response, and the way the response is delivered, in addition to what might be called the "literal content" of the response, as a source of diagnostic information.

At issue here is the role of the dialogue interaction in the diagnostic process. The rest of this chapter is broken into three sections. The first section presents evidence for the claim that interaction is a source of diagnosis; Section 2 establishes a metric, based on the data examined in Section 1, which in the future might allow an intelligent system to make use of the kinds of interactional cues utilized by human tutors. The final section of this chapter presents some concluding remarks.

INTERACTION AND DIAGNOSIS

Face-to-face tutoring consists mainly of two activities: description and explanation of some domain by the tutor, and working and solution of problems by the student. In the former, the student participates mainly by showing that he or she understands what the tutor is explaining by (a) completing the tutor's sentences, repeating the final phrases of the tutor's sentences; (b) anticipating a crucial inference from what the tutor has just said; and (c) answering occasional questions.

In the latter, the student participates by working through a collection of problems.

Both of these activities are prominent in the sessions, with varying emphasis on one or the other:[1]

Student Completes Tutor Sentence

(1)
T: the rule is (1.4) most times it k- has to be just a number, a constant If you're missing a variable
        (0.2)
⇒ S: You can't do it.
   [
T: you're in trouble.

---

[1] Saying *right* or *mhm* does not really show that the student understands what the tutor is explaining; it merely claims understanding.

(2)
T: how much energy does an electron have in a two volt field. The answer being two electron volts.
⇒ S: two electron volts.

Student Repeats Final Phrase of Tutor's Sentence

(3)
T: you know you got this three hundred electron volts, and you go- and you always go, oh my God what (0.4) what is an electron vo//lt
⇒ S: and what is an electron volt

Student Supplies Conclusion to Tutor's Statements

(4)
T: So (1.3) the main thing with the exponential function, (1.8) you have (0.2) ee to the ex (1.4) and its derivative is (2.3) ee to the ex // ( )
S: Right
    (1.7)
T: So
⇒ S: So,
    (1.6)
⇒ S: if you had ee:, to:, the:, ef of ex,
T: Right, exactly. That's (the) important case.=
S: =Right.

Student Answers Tutor Question

(5)
T: Okay, what's ohms
    (1.4)
S: It's resistance

Notice that the student's level of understanding in these passages is reflected in his or her participation in the developing conversation, rather than by simply solving steps of a problem.

Furthermore, the exact nature of the student's contribution gives the tutor an extremely delicate metric for establishing the student's level of understanding. For example, if the student completes the tutor's sentence, then that suggests one level of understanding; if the student actually anticipates that sentence, the tutor assumes an even greater level of understanding. Notice in passage (4), for example, how enthusiastic the tutor is in her response to the student's anticipation of her next move.

One can assume from these kinds of examples, then, that participation means that the student is displaying understanding; if an opportunity to perform one of these actions arises and the student remains silent, the tutor can legitimately infer from this silence that the student does not understand fully. In this way, each opportunity passed by the student is a kind of metric for indicating how well the student is keeping up with what the tutor is saying.

Consider, for instance, the following passage, in which the tutor provides an explanation for a fact she has just mentioned:

(6)
T: Okay, and here's a time (0.2) when you can drop the absolute value. (1.7) 'Kay .hh because, again as long as the inside stuff is always positive you can get rid of it, .hh and exponential functions (0.7) are always po//sitive=
S: Right, yeah.
T: =.hh so you can if you want to: drop this out, (1.5) since um exponentials (0.4) are always positive.

This passage follows a lengthy discussion concerning absolute value. If the student had understood that discussion, then he should have been able either to anticipate the tutors utterances here or complete the first one once she had started. The student does neither—he in fact remains silent—which prompts the tutor to provide an explanation for what she has just said. There are at least three points at which the student could have come in:

- After "here's a time" (note the slight pause at the end of this phrase) or somewhat later in the sentence
- During the (1.7) second silence
- At some point during the sentence beginning with "because.."

Because the student does not come in at any of these places with a turn that suggests he understands what she is getting at, the tutor continues ahead with an explanation. And, when the student finally says something, all he says is "right, yeah." which only claims understanding; it does not display it (see Schegloff, 1981, for a similar distinction). The tutor takes this as further evidence that the student's level of understanding is problematic, so she draws the "upshot" from what she has just said ("so you can if you want to drop this out"), and finally, because there is still no contribution from the student, the tutor speaks the explanation ("since um exponentials are always positive").

In the following passage, the tutor tries to elicit participation from the student; when she hears that he is not able (or willing) to join in, the tutor estimates his understanding to be insufficient for the task at hand and so offers the solution herself, laid out in detail:

(7)
T: we have one power so we'll call it (ey) over ex minus four. (0.8) .hh Okay. We want these two things to be equal (0.7) .hh so- the first thing we could do is just make them look more alike (0.7) get the common denominator. (0.3) .hh This is a step that you will skip after a few times, .hh but this is (0.2) where it comes from. Get the common denominator, so ey ex minus four plus bee ex (0.2) .hh over ex times ex minus four.

Note especially the two silences of (0.7). "We want these two things to be equal" clearly invites collaboration from the student in how to make two things equal. The tutor even offers the next step in a vernacular formulation ("make them look more alike"), followed by more silence, within which the student could offer the technical formulation. He remains silent, and the tutor offers the desired formulation ("get the common denominator").

In cases where the student is being introduced to the material for the first time, the inferences drawn from lack of active student contribution may be slightly less negative. In the fragment given here, for example, the student does not complete any tutor utterances and does not anticipate a natural conclusion, so the tutor carries out the discussion making all of the conclusions clear, providing all the propositions the student might otherwise have provided, had he understood more.

(8)
T: And now in this case all of our derivative is there (1.1) We're not missing anything .hh So you're actually just integrating one over yu de yu, .hh (1.5) .hh which gives you back a natural log. okay> .hh Um Something very nice happens (0.4) with this integration in that you actually pick up more than you (1.4) than you would expect. .hh You pick up natural log of the absolute value. (1.6) Okay? .hh (0.3) Why: is that:. (1.4) (I'll just write this out)

Timing is also used as a metric of understanding in cases where the tutor is more explicitly eliciting information from the student, as with a question. The length of time the student takes to answer, in addition of course to what he or she is doing during that time, gives the tutor subtle cues as to how easily the student is able to answer. If the student stares blankly for 1 or 2 seconds, the tutor is almost certain to intervene with a hint, or another question; if the student appears to be "working on it," the tutor is likely to give him or her leeway until either the correct answer is achieved or the student gets stuck (see chapter 5 for a detailed discussion of the mechanisms of tutor intervention).

(9)
T: so for degree two we put which degree in the numerator.
   (1.1)
T: You're down one
   (4.8)
S: Y:eah::, but this is not (0.3) factorable any further.

(10)
T: that means that- you have some other technique to do it, something that we've already learned how to do. (0.3) .hh So, what's the only other thing we know how to do. (0.8) Essentially.
   (1.1)
T: .hh You let yu equal that inside part.
   [
S: Oh, just
   (1.9)
T: Substitution

(11)
T: and what's a first degree polynomial look like
   (4.5)
S: I don't- I don't foll//ow. Just ey ex?
T: (Okay)
   (0.4)
T: Ey ex (.) plus
S: Oh, plus=
T: =Bee.=
S: =Oh, all right, yeah.

In the second activity timing in the interaction still represents a crucial source of diagnostic information, although via a slightly different mechanism than in the earlier cases. Recall that, in the second activity, the student has set about to solve problems in the textbook (in one case the student had worked through the problems before coming in and had brought in only those problems that she had been unable to solve herself).

The basic structure of these interactions is something like question–answer, only instead of a one-step answer the student must provide a step-by-step procedure for working toward the answer. The timing of these steps is a major source of information for the tutor about the student's understanding of the problem.

Early on in one of the sessions, the student and the tutor negotiate that the tutor will read the problem out loud (presumably to familiarize herself with the

74   CHAPTER 6

problem) and then the student will show what she had done previously to work the problem. But note in the following fragment that after the tutor reads the problem the student does not contribute her approach to the problem.

    (12)
    T: A one ohm wire. Okay, what's ohms
            (1.4)
    S: It's resistance
    T: Okay. Is drawn out to three times its original length. What is the resistance now.
⇒   (2.1)
    T: I think that's a hard question.
    .

    .
    T: Write down the volume of this guy.
            (1.2)
    S: Okay, how do you do volume here, let's see: (2.1) This is the area, and these are the lengths, so volume is (1.4)
    T: Area times length.
    S: Area time- oh, that's good ((laugh))

In the following fragment, the tutor helps the student through the first few steps of the problem-solving process. During the (4.8) silence, the student is trying to carry out the second step of the process, namely to factor the denominator. After a certain point in the silence, he stops moving his pencil and displays "being stuck." The student's laughter indicates that he is stuck, and the tutor intervenes.

    (13)
    T: Second thing> i:s> (1.0) to>
            (1.7)
    T: fa//ctor>
    S: factor the denominator.
    T: Right.
            (4.8)
    S: heh heh heh // heh heh, heh
    T: Okay, (0.7) here's the important thing to notice about this one. You- you copied it okay. HA HA HA

INTERACTION AS DIAGNOSTIC RESOURCE   75

The timing of the student's response (delayed with regard to their isolating of the next step), in addition to the kind of response (laughter), indicate to the tutor that the student's ability to proceed with the problem is questionable.

In passages involving student problem solving, as in the first activity, the timing of response to questions is monitored closely for clues to the student's understanding:

(14)
T: what's the difference between weight and mass.
S: I used to know this let's see
⇒   (1.7)
⇒ T: I think, (0.8) I think what it is (0.8) is that, (0.5) n- what is uh, when you do (0.6) uh:, gravity problems.=
S: =Right.//It's that-
T: What do you always do
          (1.2)
S: You have to multiply it be the-=
T: =by gee.=
S: =gee. Right.

The following passage illustrates the tutor's use of interaction to estimate how much the student understands. Just prior to the beginning of this fragment, the tutor constructed an analogy between a person crawling through two different pipes and the resistance of a wire after it has been stretched to three times its length.

(15)
T: Would it be: easier for you to go through thi(hh)s pipe, (0.9) or this pipe.
          (1.6)
T: 'Kay, 'cause we // stretched it out to three times its length.
S: Yeah.
S: Yeah.
T: It'd be a hell of a lot easier to crawl through this// pipe (than if it had) to go through this pipe.=
S: (P-)
S: =Right.

The tutor tries to elicit an answer from the student at a minimum of three places before she provides a hint; when the student still fails to initiate an answer, the tutor finally answers her own question. Notice that the student does not even collaborate in the construction of the answer once the tutor has started it, but merely claims to understand the answer once the tutor has completed it.

I would like to examine one more piece of data before looking at the metric in more detail. This fragment illustrates that the tutor responds not only to what the student says but how he or she says it:

(16)
T: Are you familiar with um> (1.6) integration by parts?
     (0.2)
S: .hh Yeah I've=
T: =Sort of.=
S: =Yea//h I've done it and, but
T: Okay.
T: Okay.

In this passage, the delay in the production of his answer, in addition to the inbreath that proceeds the answer itself, belies the student's statement. Although his answer on the surface says that he is familiar with integration by parts, his delivery of the answer says otherwise. The tutor picks up on this hedge immediately and makes it overt: "sort of" (and the student accepts this formulation).[2]

METRIC OF INTERACTION

The data presented here provide clear evidence that tutors monitor not only what students say but when and how they say it; furthermore, what is not said (and when it is not said) is just as informative to an astute tutor as what is said. How can this metric of interaction be described?

In everyday interaction when people describe something as not having happened, we take it to mean that this "something" was relevant at a particular moment, at which moment it did not take place. For instance, in July in Hawaii,

---

[2]Another framework within which one can address these issues is the preference organization (Pomerantz, 1975). According to preference organization, certain responses are socially preferred over other alternative responses. For example, if one is invited to a social event, it is socially preferred to accept the invitation and dispreferred to reject it. Similarly, if one is asked a yes–no question with positive polarity, the preferred response is yes. Furthermore, although the preferred response is usually done immediately and without much elaboration, the dispreferred response is delayed with regard to the thing it responds to, is usually accompanied by an explanation, excuse or some other kind of elaboration, and regularly starts off with a version of the preferred response.

The student's response in this passage is therefore done as a classical dispreferred: it is delayed, it has the beginnings of an explanation, and it starts with a version of the preferred response ("yeah"). The dispreferred format indicates that the student is really saying "not really" or even "no" when on the surface he is saying "yes."

it would be extremely odd for me to utter (17), since snow is not expected to occur under those conditions:

(17)
It didn't snow today.

It might be completely common, however, for me to utter the same sentence at the end of Christmas day in Boulder, Colorado. Because there is an infinite array of things that are currently not happening, when one speaks of one particular thing as not happening it must be because that thing happening was relevant at that moment (Givon, 1979; Sacks et al., 1974). So, when I speak of a student not having contributed at a particular point in the interaction, it must be because a student contribution was relevant, or expected, at that point. This research on tutoring, in conjunction with current work on everyday interaction, provides a way of approaching such expectations, and hence a metric of interaction.

Let us suppose that an interaction consists of opportunities for both parties to speak. The tutor keeps track of the opportunities that the student passes and takes and plans her own behavior using these passed and taken opportunities as a kind of metric. We need to sketch out when it is relevant for a student to contribute, or when there is an opportunity to contribute.

The most obvious slot in which a student contribution is relevant is after a question from the tutor. If the tutor asks a question, it is clearly appropriate for the student to answer. If the student remains silent in this situation, or says something that does not count as an answer to the question, we can speak of the student not answering, as having passed a relevant place to answer.

The answer is relevant even before the question is completed; if the student is keeping up with what the tutor is saying, he or she will be able to predict, after a certain portion of the tutor's question, what the rest of the question will be and when it will be possibly complete. As soon as the student "sees" in this way what the question will be asking, he or she can offer an answer. The ability to answer a question correctly before that question has even been formulated completely is a clue to the tutor that the student is understanding the material fairly well.

(18)
S: And then, the velocity would be in (5.7) uhm
    (6.0)
T: (You) think of this over here.
    (2.4)
S: Something per second
T: Right. And what's th- what's the di//stance>
⇒ S: Meters>
⇒ T: Meters, right.

Student participation is thus clearly relevant after a tutor's question. But even when the tutor does not so actively invite the student to collaborate, there are certainly places where it is relevant for the student to contribute. Following Sacks et al. (1974), I refer to these places as *transition relevance places,* inasmuch as transition to another speaker is relevant; TRP's are said to occur at the end of turn constructional units—linguistic units from word to sentence (see chapter 3 for details).

TCU=word
S: And then this is the same as that one.=
T: =Right.
       (0.3)
S: Oh: okay.

TCU=phrase
S: Not of both places, just of one plate.=
T: =Just of one plate.
S: Okay.

TCU=clause
S: 'cause I don't think I had a dielectric (0.6) wait, no, I do:
       (0.9)
T: Ts, 'cause you have paraffin.

TCU=sentence
S: I just assumed that one was positive and one was negative.=
T: =Yeah, that's the only way it makes sense.

This system works as follows in most interaction: Each speaker is allotted one TCU; at the end of a TCU by Person A, Person B may speak; if B chooses not to speak, A may speak again.

This normal turn-taking system can be temporarily suspended, however, so that one speaker can take what is called a multi-unit turn, for example to tell a story or explain a complicated aspect of differentiation. This suspension does not imply that no one else can speak at all; the recipient of the multi-unit turn nods, makes comments and in general collaborates in the production of the turn, these productions being located at or near the end of clause boundaries. The clause boundary, and especially the sentence boundary, of a multi-unit turn from the tutor is thus a relevant place for students to contribute.

The approach of the end of a clause and/or sentence boundary in a multi-unit turn raises several opportunities for the student, as we saw in the last section. The student can (a) complete the current clause, (b) repeat the end of the clause once it has come to a possible completion, (c) anticipate the tutor's next clause/sentence, or (d) ask a relevant question. These can be ranked in order of cognitive difficulty 3 > 4 > 1 > 2; the option chosen by the student, if he or she opts to do more than a simple continuer (e.g., *mhm, right*), gives the tutor some indication of how well the student is following.

A careful examination of the transcripts reveals that students insert their utterances within a multi-unit tutor turn within a span of relevance (the transition relevance place), lasting roughly from the last constituent of clause (a) to the first constituent of clause (b), as shown schematically here:

---------------/----. --/----------

Within this span, student contribution is relevant. In fact, it is relevant at two different structural slots within the span, namely within Clause a and in the transition space between Clause a and b. Because within this span of relevance active participation from the student is relevant, when the student does not say anything, or produces only a continuer, then the tutor can legitimately infer certain things about the student's understanding.

I turn now to a case in which the student is the main speaker.

Within the activity of a student solving a problem, the student verbalizes steps of the procedure to display his or her understanding of each step. In some cases the student will produce a sentence that displays a correct understanding of one step, or of a preliminary to a step, and then will fail to produce the step that must logically follow, indicating that he or she does not see how the step he or she did produce is a resource for continuing the rest of the problem.

(19)
⇒ S: Ey plus cee equals zero.
⇒   (0.3)
⇒ T: Right, so that tells // you
    S: Ey equals cee.
         (0.4)
    T: Minus cee.

(20)
⇒ S: So it's got to be: in our fourth quadrant.
    T: Right.
⇒   (1.4)
⇒ S: Aha. (0.8) Aha, o//kay.
⇒ T: Which means?

S: Then I have to come back down here and I'm- you're asking me to choose a sign, righ//t?
T: Right
S: .hh Okay.
T: For your cosine.

In each of these cases a correct preliminary move is followed by a bit of silence, wherein the tutor hears that the student is not going to produce the substantive move that logically follows from the preliminary move. Here again, a further student contribution is relevant, and when it is not forthcoming, the tutor overtly solicits it.

We thus have three foci within which a metric of interaction for tutoring is necessary: (a) after a tutor question, (b) in the span of opportunity within a multi-unit tutor turn, and (c) after a student utterance whose implication for the next step of solving a problem has not been elaborated. Of course, a tutor can always elicit student participation in other contexts, using other devices—the laughter in passage (15), for example (see Jefferson, 1979), for a discussion of laughter-inviting participation)—but these are the most common spaces of student relevance.

CONCLUSIONS

The goal of this chapter has been to demonstrate that tutor's base their assessment of student understanding on more than steps in a problem-solving task, and to describe how assessment is achieved.

Although I have focused in this chapter on the tutor's use of timing in the interaction, it is worth pointing out here that from the student's perspective there is also a metric of tutor response. If the tutor starts nodding and agreeing with the student's turn even before that turn is possibly complete, then the student can infer that the answer displays a high level of understanding, whereas if the tutor delays responding after a student's turn, then the student will hear that to mean that the last turn was in some way less than fully correct or appropriate. The following fragments illustrate the importance of the timing of the tutor's contributions.

(21)
[T and S are talking about absolute value]
T: if ex (.) were negative to start with (0.6) now minus ex would be positive. .hh So:=
S: =Oh so you have to ah (1.2) know what the number is befo//re you can say
T: Right, exactly. Before you can simplify.

In (21), the tutor overlaps the student, producing an "early" assessment, that is one that precedes a transition relevance place. The earliness of the tutor's response correlates with its extremely positive nature (*right, exactly*).

In the following fragment, the "lateness" of the tutor's contribution is significant for the student, in that it correlates with the negative assessment of that contribution:

(22)
S: That should be twelve shouldn't it?
    (1.0)
T: No: I think it should be, twenty four.

Thus, both student and tutor carefully monitor the timing of the interaction.

This work treats all dialogue as deeply interactional. This perspective includes thinking of interaction not only as what people do say, but what they don't say, and when they don't say it, that is, as passed and taken opportunities; and thinking of what appears to be monologic material, such as long tutor explanations, as in fact interactional, produced in delicate response to minute facts about the recipient's behavior. Tutoring interactions are not treated as pre-planned, because even in the most monologic situations the talk is still fundamentally interactional, responding as it does to things said and not said by the recipient, things whose presence or absence could not have been predicted in advance.

This chapter has explored the use of timing in the tutoring sessions. In the next chapter, we examine the closing of sessions and the ending of segments of sessions.

7

The Target of Tutoring

GOAL OF THE CHAPTER

This chapter explores the closings of our tutoring sessions, as well as the endings of sequences within each session. In particular, this chapter focuses on the processes by which the tutor and student negotiate what is going to count as "the answer" to a particular problem they have been working on. The findings presented here reinforce the larger hypotheses of this study regarding the contextualization of abstract forms and the local management of plans.

The negotiation of answers is relevant to ITS design in part because of current literature on student modeling. Wilkins, Clancey, and Buchanan (1988) describe most tutoring systems as utilizing differential modeling, the process of: "identifying differences between the observed behavior of a problem-solving agent and the behavior that would be expected in accordance with an explicit model of problem solving" (p. 258).

The explicit model of problem solving is in many systems known as the expert module, or expert system (Anderson, 1988). Expert modules vary in the degree to which they try to capture human expert knowledge: "Black box" modules generate correct behavior in solving a problem, but the computational methods by which they do so is opaque to the user and not based on human methods; "glass box" modules are constructed using knowledge engineering techniques with the help of a human domain expert, but the processes in which they engage their knowledge are not characteristic of human processing; "cognitive models" aim to simulate human problem solving using human knowledge strategies and human methods of knowledge representation (Anderson, 1988). In cognitive models, procedural knowledge tends to be captured in if–then rule systems (typically production rules), and declarative knowledge tends to be represented with semantic nets (Bobrow & Winograd, 1977) and schema systems (Schank & Abelson, 1977).

Because students are seen as moving eventually from novice to expert status, the student and expert models are typically constructed with the same system of knowledge representation (VanLehn, 1988); and because of the difficulty of building two entirely different models for student and expert, many systems make use of essentially one model: "economy and other implementation considerations frequently dictate a merger of the two models. The student model is represented as the expert model plus a collection of differences" (VanLehn, 1988, p. 62).

Although conceptually the tutor and student models could diverge, then, in practice one is often treated as an errorful version of the other. In this way tutoring is seen as a process of eliminating differences between the student and the expert model (see also many of the contributions in Sleeman & Brown, 1982).

The notion that learning takes one from a more novice-like state to a more expert-like state is unarguable, but the view of tutoring as identifying and eliminating differences between tutor and student is worth questioning. I argue that the immediate goal of tutoring is not best thought of as expert knowledge, or even some abstract notion of mastery, and that the process of tutoring is not best conceived of as just eliminating differences between a student and an expert model. Rather, the goal of a particular session, or segment of a session, is inextricably tied to particulars of the situation.

THE TARGET OF TUTORING

In this section we explore passages from face-to-face tutoring sessions.

The first and most obvious indication that the target of tutoring is worked out for a particular occasion of problem solving can be found in passages like the following, in which the tutor makes overt reference to the fact that certain problem-solving strategies would be helpful for the student but that those strategies are somehow "beyond" the student at this point:[1]

(1)
T: And what I recommend doing ((clears throat)) is once you think you have the numbers, (0.9) plug them in and check it, (0.7) because I usually have to (0.4) go back like two or three times, to catch all my little

---

[1]Even the intended definition of the term *expert* is unclear: VanLehn (1988) defined it as mastery; many others offer no definition at all.

mistakes. If you know how to use matrices, and I guess you don't learn that til fourth semester.

(0.9)

S: Ah we yeah, that's mostly, where you learn, in differential equations we learn about matrices, but we- we've had some matrices already.

T: They help. Because they get rid of all these equations floating around your page and they just (0.2) put them all together and you don't have to look at the variables any more

[they continue working, without matrices]

If the tutor's only goal is to bring the student's understanding and abilities in line with her own, then she would have taught the student how to solve the problem using matrices. Instead, she chooses a strategy that is more in keeping with her sense of what the student is capable of at the moment.

This example suggests that tutors do not use their own knowledge as the standard against which student performance is evaluated; rather, they use some sense of what this particular student needs and is capable of right now.

The second indication that the target of a particular tutoring session, segment of a session, is context-dependent can be seen in examples like the following, in which the tutor makes use of the student's real-world situation (a test the next day) in formulating her contributions. Because the tutor herself is not taking the test, and because presumably the tutor's knowledge is not organized toward a goal of test-taking, or not organized only toward this goal, this phenomenon cannot be interpreted as guiding the student toward an abstract expert understanding; rather, the tutor formulates her suggestions in a manner which services the particular needs of this particular student on this particular occasion.

(2)

T: So the thing that I would do is on tests and stuff (even though like) you know all the numbers?

S: Aha

T: You know? And I still do this, is I writ- I wri- I like (want to) plug in the numbers and get (the) answer half way down.

S: Aha

T: But the thing is is just deal with the letters, (0.5) and the symbols. Because for instance you might have, in this case you don't, but let's say you might have the mass, of one of these things on both sides of the equation. Let's say there was a mass in here?

S: Aha

T: And they would cancel out.

S: Right.

T: And you would never need to put that number in.
S: So do all the algebra first?
T: Exactly.

Both of these cases suggest that tutors do not guide students toward a single "correct" set of solution steps and answer, or even toward an expert-like understanding separate from the students' own needs and abilities.

In a similar vein, when students try to produce what might be characterized as "expert" behavior (i.e., the way an expert would approach a problem), they are often chastised by their tutors for producing behavior that hides what the students do not know and that could lead to errors. In other words, it seems that expert behavior is considered "dangerous" for students under certain circumstances:

(3)
⇒ T: Now you did too many steps at once, I can't figure what you did
S: Oh ((laugh))
T: ((laugh))
S: Okay. (0.6) This is my equation. So I just (1.3) put r over there, and and then // you divide
T: You had an mg here. (0.4) Okay
S: You divide by f. I switched f and mg.
T: Good. Now one thing (0.4) on a test, (1.0) like I did know what you did here, but my point was like let's say you're on a test, you might be freaking out, right?
S: Right, right.
T: And it's really easy to put the wrong thing on the top or the bottom.
S: Right.

The examples just provided illustrate, in an obvious way, that the attainment of expert knowledge–and in particular a single, context-independent model of expert knowledge—is not the only lighthouse for tutoring. The following kinds of events, however, indicate some not so obvious ways in which this is the case.

As Lave (1988) demonstrated for adult match users, the setting in which math activity takes place and the math activity itself are mutually constitutive of each other, that is, they bring one another into being:

> If given a "going to the store" problem to do in a math class most people would treat the story as having no substantive significance—it is there to disguise mathematical relations. The same people generating math dilemmas in the supermarket are likely to organize quantitative relations to fit the issues and concerns of buying food. . . . Neither math nor shopping would be organized in the same fashion across the two situations. The proportional contribution of each to the process of

activity as a whole varies from one occasion to the other, there is no fixed procedure for math or shopping . . . (p. 99)

In our case, what this means is that the room in which the tutoring takes place, the book(s) the participants use, their goals and real-world tasks (taking exams, being finished in 1 hour, etc.), all constitute what will count as a problem and what will count as an appropriate solution for a problem; whereas the chosen problem and the interpretation of that problem shed new meaning on taking a test tomorrow, having an hour to work together, and so on. In this view, it is overly simplistic to think of tutoring as a process that moves a student toward a single, "correct," expert understanding.

The following passage provides a clear illustration of the mutually constitutive nature of activity and setting (tutoring and context). In this fragment, the tutor and student negotiate together what will count as having completed a particular physics problem. There is no fixed level of solution—corresponding to what an abstract expert might require—toward which the pair works; they work together to decide when they have done enough:

(4)
T: You can say that's zero without doing all this square root st//uff.
S: Aha
T: Right? So you always kind of want to just to watch the symbols and just // go wow, you know, this is (1.0) working out.
S: Aha
T: Okay, so ah, let's figure this out, this will be interesting.
    (1.3)
S: Okay (0.7) you mean plug in the numbers?
T: Yeah.
    (0.4)
S: Okay. So I need to find (5.8) the force (2.1) it's equal to um (3.5) it's not the force, it's electric field, kay, kyu (0.6) mi//nus
T: Oh no no no. Don't fi- I mean let's don't- oh yeah, I guess you have to figure all that out. Okay, let's don't figure it out then.
S: Okay, ((laugh))
T: I think you're pretty confident.

Here the activity of tutoring on a given problem is shaped by the situation of (a) having only 1 hour for the session, (b) getting to a stage in the solution that is no longer physics per se (arithmetic), (c) getting to a stage from which "any calculator" could figure out the answer, and (d) getting to a kind of calculation (arithmetic) that the student has demonstrated she can manage.

Thus, the history of the particular tutorial dialogue, the knowledge of the student, and the general goals of the interaction all help to constrain what counts as an adequate solution for this particular problem.

In the next passage, the same tutor makes the context-dependence of what counts as a "good" solution overt:

(5)
T: like (0.5) if I'm grading these exams if someone writes down the right equation, okay
S: Right.
T: With all the right things. And then makes (the) mistake on plugging the numbers in, that's not that bad of a mistake, right?

Here, the test-taking context of this particular tutoring session very much influences the tutor and student's notion of what will count as a good or correct solution.

This view of tutoring leads to a conceptual revision of certain aspects of the tutoring process, including the statement of the problem to be solved, and the tutor's role in interpreting that statement. From this perspective, the statement of the problem (e.g., as one would find in a textbook) does not describe—out of context—how to solve the problem; it provides a system of soft constraints that any solution must take account of (for a discussion of the underspecification of instructions, see Suchman, 1987). One of the jobs of the tutor and student, then, is to contextualize the problem statement so that it can be worked at in a particular setting, for particular goals, and so on.

In the following fragment, the tutor and student are working on a problem that ends up requiring the student to make use of the notion of potential energy. The student expresses confusion about this concept: When she asked her teaching assistant about it, the TA said "they take potential energy to be 0 at infinity." The student now is stuck with turning this statement into one she can use procedurally for a given problem:

(6)
S: That's- that's what I don't understand. I thought (0.7) in reading the chapter and listening to his lectures and stuff, (1.0) that (0.8) potential energy, could only be measured as a- a change.
    (0.8)
T: Aha.
S: Or something. And then I asked my TA, briefly during our rec- recitation this morning and he said oh, well they take (0.8) potential energy to be zero at infinity

T: Right
S: So you use that. And I don't know really how to (0.2) plug that in.

The tutor and the student now work together with an equation for potential energy to demonstrate to the student that zero at infinity arises out of the relationships in the problem—it is not a procedural description of potential energy:

(7)
S: So actually when I'm calculating this. I don't have to add in that it's zero at infinity that's just
    (1.0)
T: No.
S: I just
T: Because see look, (0.6) let's look at this equation here. What happens when r gets really large.
    (1.0)
S: Then the force on this gets weaker and weaker.
    (0.7)
T: Or yeah. But just this guy here.
S: Yeah.
T: And you've got this field out here.
S: Aha.
T: But you make this vector r so large.
S: he can't
    (0.8)
T: That r is infinity and this goes to zero. So in other words if you put this little q, (0.5) way out at infinity. Then the potential energy that he would have, due to this charge, would be nothing.

Thus, although zero at infinity could be taken as an expert's understanding of potential energy, it is not an understanding that will help this particular student solve this particular problem; the tutor's suggestion that they understand zero at infinity by looking at the equation for potential energy provides the student with a contextually appropriate formulation, one that interprets the otherwise opaque formula.

As discussed in chapter 1, we found, in fact, that one of the major tasks of our tutors was the contextualization of problem statements. The tutors reformulated problem statements so that their similarity to problems previously solved by the student would be apparent, they taught the procedure of "thinking of what the answer should be" before the solution was even started; they taught students to see what it "really was" that the problem asked for, or what a number, which might otherwise seem arbitrary, "really meant." That is, they worked not to propel the students through a prespecified series of solution steps, but rather to create a meaningful problem, where meaningfulness is constituted by the needs and abilities of the tutor and student in that particular situation—with an immediately recognizable solution path. Because the meaning of a problem depended on the individual student and tutor, and the history of the dialogue (among other things), there was no single correct understanding of a problem statement, and no single correct solution path. The following passage from Lave (1988) illustrates the context-dependency of problem interpretation and solution path:

> Another problem posed to new members of Weight Watchers in their kitchens provides a further illustration. As in the exercise concerning peanut butter sandwiches, the dieters were asked to prepare their lunch to meet specifications laid out by the observer. In this case they were to fix a serving of cottage cheese, supposing that the amount allotted for the meal was three-quarters of the two-thirds cup the program allowed. The problem solver in this example began the task muttering that he had taken a calculus course in college. . . . Then after a pause he suddenly announced that he had "got it!" From then on he appeared certain he was correct, even before carrying out the procedure. He filled a measuring cup two-thirds full of cottage cheese, dumped it out on a cutting board, patted it into a circle, marked a cross on it, scooped away one quadrant, and served the rest. (p. 165)

Lave commented "Thus, "take three quarters of two-thirds of a cup of cottage cheese" was not just the problem statement but also the solution to the problem and the procedure for solving it" (p. 165).

In this example, the subject works hard to make the problem statement meaningful to the situation that he is in—a kitchen, with measuring cups and a container of cottage cheese. Our tutoring examples are not quite as extreme as this behavior, but they show similar patterns. In the following passage, for instance, the tutor tries to get the student to see the solution in the problem statement, as reflected in the student's drawing of the problem:

(8)
⇒ T: Okay now, before you start. What should your answer be.
    (0.7)
T: In what direction. Like is it
S: In what direction?
T: Yeah.
    (2.1)
S: It'd be like in that direction?
T: Okay now why do you say that.
    (2.5)

S: Oh wait (1.5) this one is farther away. I th- I was thinking that it's the same distance. This one's closer, so it's going to be more
T: Okay.
  (0.9)
T: This is going to be stronger.

On the surface it seems bizarre to ask a student what the answer should be before the student has even started to solve the problem, if one sees working through a problem as a unidirectional process. But as Lave (1988) showed for adult math users in everyday settings, solving a problem is an iterative process, which involves a procedure Lave called "gap-closing." By gap-closing she meant that a given interpretation of a problem statement is in part produced by an initial idea of what the solution should look like, and that initial idea may be modified (or the interpretation of the problem statement may be modified) as the solution progresses.

Another example of the tutor actively teaching the procedure of gap-closing follows:

(9)
T: Now the way- let's say y- so you're multiplying this 350 times a very small number.
S: Aha
T: And that makes sense, because joules, joules are units that you use in every day life.
S: Aha
T: And electrons are really small. And the unit that's good for electrons is electron volts, which is a very small amount of energy. So you should make sure when you get your answer, that you don't get (0.6) 350 electron volts isn't 9 million joules.
S: Right. // It should be smaller.
T: It should be a very small number of joules.

Here the tutor encourages the student to check her answer intuitively, at a gross level, to make sure that it fits the expected solution.

Further evidence that the activity of solving a problem and the setting of this particular kind of tutoring mutually constitute one another can be seen in the following passage:

(10)
T: When you look at this, you know you got this 350 electron volts, and you go- and you always go, oh my god what (0.4) what is an electron volt
S: and what is an electron volt

T: And then if you can any way fool around with the units, to figure out well what is it I'm talking about you know, and you go well it's q times v and you look over (at) your equation and you go okay well that's (0.2) work and energy are the same thing.
S: Aha
T: And then you go, energy, okay, well what energy equation do I know that has velocit//y in it.
S: s: velocity

Notice that in this passage, which occurs after the tutor and the student have worked their way through most of a problem, the tutor is actually teaching the student how to see what the problem statement is "really" asking for ("what is it I'm talking about"). What is fascinating about this passage is that the tutor enacts for the student the exact mental processes that the student should go through in order to arrive at the interpretation of the problem statement. But notice that these are not the mental processes of an expert; this is not really how the tutor would talk to herself in solving a problem of this kind. She would have this process compiled for a problem of this level and would not need to go through the entire step-by-step procedure. Nonetheless, although the procedure is not one an expert would grind through, it is appropriate for this particular student, who had tried to solve this problem on her own and could not begin to figure out what the problem statement was asking for.

Our tutors acted as translators by putting the problem statement into terms meaningful to the individual student:

(11)
T: (reading the problem statement) How strong is the electric field between the plates of a twenty microfarad.
⇒ Now what's microfarads, (//   )
S: Ten to the negative six
    (0.7)
T: Okay, so what is a farad, what units is that.
    (0.6)
S: Wait- oh what's a farad equal?
T: Like no, I uh- are they talking about length? That's what I want (to know).
S: Oh. (0.6) Uhm, the farad is the capacitance.
T: Good.

T: Okay. A twenty microfarad, so this is a very small amount of capacita//nce, right

S: Mhm
   (0.3)
⇒ T: Air gap capacitor. So, air gap means there's air between.
S: Mhm
T: If they are two millimeters apart and each has a charge of 300 (0.8) micro (0.6) coulombs. Okay.

In interpreting the problem statement, the tutor can also make use of the history of their dialogue to classify a problem as "exactly like" a problem they have already solved, "identical" except in one small feature, or "new." Because specific information about how to solve the problem is not usually given in the problem statement, such information is nevertheless crucial for students who are learning which changes in features count as significant and which are merely "disguising" similarity:

(12)
T: (reading from a sample exam) Given three point charges, fixed at three corners of a square. Find the electric field and tendency vector e, at the corner pee, with no charge, this is exactly like the problem we just did.
S: Right.

Given the surface features of this problem, it is not obvious that its solution path will be identical to the problem they have just worked ("Electric field due to positive charges shown. Calculate magnitude and direction, is your result consistent."). As in any concept formation task, the novice must be guided to know, as experts know, what problems have the same "underlying structure" and hence similar procedures for solution. What makes this particular example unlike teaching expert knowledge, however, is that there is no overt attempt to teach the features that are crucially the same—the student is left to discern what makes the second problem "exactly like" the first problem. Moreover, the student may abstract nothing more general than what holds those two problems together as a category; she may not generalize, presumably as an expert does, to entire ranges of other problems (including via analogy in some cases). That is, what the student is becoming good at is student knowledge, or what a student at this level is expected to know, in particular what this particular student, in this particular situation, at this level is expected to know.

CONCLUSION

This chapter has tried to demonstrate that the target of tutoring—the solution path taken, as well as the end solution—are dynamically negotiable, being constituted by aspects of the setting (including participants' goals, physical location, time period, history of the dialogue, etc.), and do not necessarily reflect a single notion of mastery. This is not to say that concerns of mastery are not attended to in these tutoring sessions; on the contrary, the tutors were clearly motivated to help the students improve their understanding and performance, both obviously issues of mastery. But two critical points must be borne in mind: first, more is being done in these sessions than eliminating bugs and misconceptions. Students are learning new procedures and strategies for interpreting problem statements, gap-closing techniques, test-taking strategies, and so on; Second, what counts as mastery for a given problem or session, and how that level of mastery is achieved, cannot be stated in advance of the session, but rather must be viewed as an *achieved outcome* of the interactional work of the two participants.

One final point is in order. Lave's findings that math activity is shaped by the setting in which it takes place (and which it also shapes) was perhaps not surprising given the settings she examined—supermarkets and kitchens—where accuracy and correct solutions are not crucial for making decisions. In fact, in many of Lave's examples, her supermarket subjects correctly calculated price per unit for competing brands of a given item and then deliberately chose the more expensive brand, claiming that they liked the taste better, or that it was packaged in a more convenient size.

In our tutoring sessions, on the other hand, one would have expected that both solution path and solution would have been completely specified in advance and would not have been shaped by such real-world concerns as what grade the student wanted in the tutored-for class, or whether the tutored-for exam was going to be multiple choice or problem solving. But in fact, these kinds of concerns were regularly integrated into the tutoring process, as were concerns of time, history of the dialogue, student's background in the subject matter, student's self-diagnosed style of learning, and so forth. Lave's conclusions then hold not only for such settings as supermarkets and kitchens: They seem to hold for settings like physics tutoring as well. In that all human activity—no matter how isolated and cerebral it may seem—is situated, it is *always* the case that activity and setting are mutually constitutive.

8 Bandwidth

PRELIMINARIES

As part of the larger project on tutoring, we contrasted two different modes of communication: face-to-face and terminal-to-terminal (see chapter 2 for a detailed description). Because of its deeper theoretical interest, most of the present study has focused on properties of the former. But the latter turned out to provide some startling insights into the nature of tutoring, and into communication within a narrow bandwidth; this chapter therefore reports on some of the findings of the bandwidth comparison.

TURN-TAKING AND REPAIR

One of the little understood and equally little studied aspects of natural language interface design lies in the area of turn-taking between user and system.[1] In many systems the user is allowed a single command at a time, whereas the system is allowed a large section of canned text; in others, the student can be interrupted immediately after producing some kind of error, but is given no facilities for interrupting the system. These various solutions to the turn-taking problem ap-

---

[1]To give an idea of how little studied this area is, I found scant mention of this topic in the interface design literature. See Coombs and Alty (1980); Gaines and Shaw (1986); Baldwin and Siklossy (1977); Miller and Thomas (1977); Fitter (1979); Kidd and Cooper (1985); Edmonds (1982); the issue of *IJMMS* edited by Rissland; Hayes and Reddy (1983). Of course, turn-taking, and repair, are brought up as central notions in Suchman (1987), and in the quote from Seely Brown given in the preface.

pear to be unprincipled and ad hoc, designed to satisfy the particular needs of a particular system, at least as those needs are understood by the designer of the system.

In this section, I examine, in a principled fashion, the turn-taking system required for interactive systems. The findings reported here strongly suggest that the need for repair of one's own and one's interlocutor's utterances critically affects the kind of turn-taking mechanism that will be viable in a tutoring system.

Results

One of the most striking findings of this study was manifested in numerous complaints from tutors and students that they had made a mistake somewhere in their text, not noticed it until after they had sent the text off, and had been unable to fix it until after the other person had replied, thus making it possible for the error to confuse the hearer. This was the one universally agreed upon design flaw of the system.

It is clear from these findings that any kind of communication model, including a computer interface, must have built into it a reasonable and effective mechanism for fixing trouble in the communication when such trouble arises. Exactly how such a mechanism should function is the topic of the remainder of this section.

Turn-Taking and Repair in Face-to-Face Tutoring

The face-to-face tutoring sessions we videotaped exhibited a much different turn-taking system from the one described earlier for the terminal-to-terminal session. In face-to-face tutoring, as in everyday conversation, turns are allocated by a much more flexible system: Speaker A can speak again even if there has been no overt response from Speaker B, and either participant can interrupt the other—in the middle of a sentence or during a longer turn—to repair a misunderstanding (or can overtly show that no repair is needed, for example by saying "mhm," "yes," "right," etc.). One possible mechanism of turn-taking that could generate this kind of flexibility is presented in chapter 3. The basic rules are repeated here for convenience:[2]

Rule 1 applies initially at the first TRP of any turn.

(a) If the current speaker selects a next speaker in current turn, then current speaker must stop speaking, and that next speaker must speak next, transition occurring at the first TRP after next-speaker selection.
(b) If current speaker does not select next speaker, then any (other) party may self-select, first speaker gaining rights to the next turn.

---

[2]For a discussion of the nondeterminative nature of these rules, see chapter 3.

(c) If current speaker has not selected a next speaker, and no other party self-selects under option (b), then current speaker may (but need not) continue.

Rule 2 applies at all subsequent TRP's. When Rule 1(c) has been applied by the current speaker, then at the next TRP Rules 1(a)–(c) apply, and recursively at the next TRP, until speaker change is effected.

This system also accounts for a number of other interesting characteristics of the face-to-face tutoring data collected. For example, one of the striking characteristics of these data is the variability of turn length: In some cases, tutors pursued extremely long, multi-unit turns, whereas in others they constructed very short turns; similarly, in somewhat rarer cases students produced long turns, whereas other times they produced short, one-unit turns. Longer tutor turns seemed to correlate with a teaching style, but even in a highly interactive, nonteaching session, one can find an occasional long turn from the tutor. Examples illustrating this variability are given here:

(1)
S: And uh (0.6) And okay, so, (0.5) uhm (1.3) and I have a value of three, and they want to know what the tangent is. So, I have one over cosine of theta equals three. (0.8) And I have the sine of theta over cosine of theta (1.0) u:hm. (0.8) .hh Okay, so I guess- I somehow have to: (0.8) tangent of theta is going to be: (0.4) sine of theta over cosine of theta. (2.0) One over cosine of theta

(2)
T: You know their ratio
S: Yeah.
T: But you don't know exactly what they are.
S: What they are, yeah.
T: So=
S: =Okay. So this is the basic.
T: Yeah.

(3)
T: The outside derivatives get one over this inside function, times the inside derivative. .hh So .hh this actually gives us a real powerful integration tool. Surprisingly powerful. .hh Because now if we have ( ) some function like this (1.2) and we want to differentiate it, .hh (.) the natural log disappears in the differentiation process. You take your inside derivative, (0.5) three ex squared plus six, divided by the inside function, (.) by the chain rule, .hh and then get a rational function.

Clearly, a turn-taking system like the one presented here is needed to produce flexibility in turn length, with no single party distributing turns. It is not the case that all tutor or student turns are long, or that all possibly lengthy turns are completed (in many cases they are interrupted by the other participant), but rather, that length of turn is freely negotiated between speaker and hearer, as it is in everyday conversation, and not predetermined by a rigid on–off kind of mechanism. The only kind of mechanism that could produce such flexibility in turn structure is one that allows: (a) speakers to speak again, even if the interlocutor has not in the meantime taken a turn; (b) participants to repair a previous utterance, even if it looked as though that turn was already complete; (c) participants to speak before the other's turn is complete, thereby allowing the hearer to interrupt the speaker in the middle of a sentence or at the ends of units in a long turn.

An additional prominent characteristic of the face-to-face data is the lack of tutor control over turn allocation. This lack of control contrasts with the typical classroom setting, in which the teacher has control over who speaks, for how long, and when. Turn-taking in the classroom is modulated in the following way: The teacher asks a question, the students then raise their hands, forming a pool of possible next speakers, and the teacher selects one person to speak next. After that person is finished speaking, the turn bounces back to the teacher, rather than to one of the other students (this system is described in some detail in Schegloff, 1987). In our tutoring data, no such turn allocation exhibits itself.

Moreover, the tutoring turn-taking system provides for a fundamental process of tutoring: Correction. This point can be nicely illustrated by contrasting the possibilities for correction in face-to-face tutoring with the possibilities for correction in the terminal-to-terminal tutoring I described earlier. The terminal-to-terminal interface allowed for multi-unit turns (in fact encouraged them) but did not allow for two critical elements: (a) the opportunity for the hearer to indicate understanding or lack of understanding at the end of every unit; and (b) the opportunity for the speaker to initiate correction on his or her own turn after it was sent. The problems with this method of turn-taking are immediately apparent. For instance, if the tutor constructs a 20-line text with an error (which goes unnoticed) in the first line, then the student receives this text and can't understand more than the first line, because the error completely confuses the explanation the tutor meant to be conveying. The student then sends a repair initiator to the tutor, who then must reconstruct in her mind what the error could have been, and how it affected the rest of the text. Then the tutor must repair the error and reproduce the other 19 lines of text she had already sent. If this happened with every turn, the efficiency of the communication would drop close to zero and very little learning would take place. In practice, participants in the terminal-to-terminal environment quickly learned to take relatively short turns when possible; that is, they tried to manipulate the interface they were given to make it match more closely the turn-taking system of natural conversation.

In contrast, the turn-taking system of face-to-face tutoring enables speakers to produce multi-unit turns and allows the hearer to initiate repair at the end of every unit. Thus, the speaker can initiate repair, even after the end of the multi-unit turn, if there is some sign, including silence, that the hearer is having difficulty understanding.

The fundamental principle underlying this system is that each speaker is allotted one unit (TCU) per turn; at the end of such a unit, someone else can begin to speak. But, we know that even in natural conversation people sometimes take very long turns—for example, stories, jokes, reports (see, e.g., Schegloff, 1981). Such multi-unit turns, however, regularly display to the hearer at the outset that something long is in store (e.g., "guess what," "you'll never guess what happened to me today," "did you hear the one about . . ."). So we know that even in a system that allocates one unit per turn, there are ways to secure multi-unit turns (see Sacks, 1992).

But, securing a multi-unit turn in the course of a tutoring session does not mean, as one might expect, that one person speaks and the other is silent. Although a speaker may have secured a multi-unit turn, the hearer will produce things like "uh huh," "mhm," at predictable points within the multi-unit turn.[3] In fact, these so-called backchannel utterances are positioned at the end of TCUs and signal that, given the end of a new TCU, the hearer is having no difficulty understanding or agreeing with what the speaker has said. In other words, even in a multi-unit turn, there are places—namely the end of each TCU within the larger turn—where the hearer can initiate repair ("huh," "what," "I don't get it," and the like). In the following passage, S, the hearer of a multi-unit turn, indicates at many TCU boundaries that she is understanding and agreeing, and not about to initiate correction or repair.

(4)
T: Let's say you have a triangle, (1.2) and this angle is theta, and this is ar.=
      [
⇒ S: Aha
⇒ S: =Right.=
T: =Okay? (0.7) This'd be the wy. Wy equals ar sine theta. And ex, equals ar cosine theta.
      [
⇒ S: ar cosine theta. Right.
T: And- and I use that all the time, it has to be something I've used like (0.4) over and over and over.=
⇒ S: =Mhm.

---

[3]This seems to contradict Hayes and Reddy's (1983) notion of *implicit confirmation,* wherein no overt indication of comprehension is necessary.

Without a facility for producing such backchannel comments, the student can become confused because there is no opportunity for the student (or the tutor, in the case of a long student turn) to clear up misunderstandings as they occur. Thus, misunderstandings may accumulate, so that repair becomes extremely difficult and time-consuming (see Anderson et al., 1987, for a similar point). However, if the level of understanding and agreement is constantly being displayed by both parties, then there is little opportunity for serious misunderstandings to arise.

Moreover, this agreement about the level of understanding is facilitated by mechanisms for interruption, as a device for repair (as well as to display understanding, that is, no need for repair). There are instances in conversation where one participant starts talking before the other participant has finished a complete unit, usually to initiate or perform some kind of repair. Although this facility is not captured well by the rules previously given, it is clear from looking at the tutoring data that certain kinds of interruption are essential for maintaining mutual comprehension.

In the following passage, for example, the tutor interruption repairs the misunderstanding that has grown because the two have been looking at different problems.

(5)
S: And (0.5) the magnitude would be:
   (13.5)
S: The equations I have f- f- for this is uhm, force over charge>
   (1.6)
S: But (1.5) I don't know if force would be (//)
T: Okay, nah, nah you've got another one (0.9) You have one that says down there, energy stored in capacitor.
S: O:h.
   (3.9)
S: Energy stored in capac- but- don't they want
   (3.5)
S: So this is electric field? (0.9) This is e-
⇒ T: Oh, excuse me, excuse me, I was looking at this one, and you're still working on this.

In the following passage, the tutor interrupts the student to prevent a potential difficulty.

(6)
S: Equal to one half. And then the mass. (0.8) I need the mass of an electron.

T: Mhm
      (1.8)
S: And that's in my book. (1.7) and th//en
⇒ T: e-e-e- what units are you going to put that in. This is the main thing I'm worried about.

In the following case, the student interrupts the tutor to clarify the significance of something the tutor has suggested:

(7)
T: Does that make sense to you, or not //yet.
S: Well to- not- I mean, (no sense really).=
T: Okay. Well (0.4) just start (0.6) like for example=
⇒ S: =It seems like that's just you know the- the- back of the problems

Both aspects of the design of turn-taking in tutoring function to enhance mutual understanding, by allowing misunderstandings to be clarified as soon as they come up (in some cases, even before there are overt symptoms of misunderstanding). Both participants can remember what is likely to need repair, without having to repeat long portions of previous turns.

In addition to providing for on-line repair, this turn-taking system also allows the speaker to speak again without waiting for the hearer to complete a turn first. For example, in the following passages, the speaker talks again even though the hearer has not completed a turn in between:

(8)
T: And you probably have some way of remembering that already.
      (1.3) [student stares blankly at tutor]
T: Maybe not>

(9)
T: So one electron volt is the charge of an electron, which is, one point six times ten to the minus nineteen joules.
      (1.7)
T: So I don't know, do you have to write that number down?
      (1.1) [student stares blankly at tutor]
T: Can you write that down on your piece of paper?
S: Sure

In these passages, the tutor expects a confirmation of or answer to her first line; the student, however, displays signs of not being able to provide the response, so the tutor speaks again.

Further Refinements

There has been at least one tutoring system developed (see Anderson et al., 1987) that has built into it a turn-taking system that in one respect matches the suggestions I have given for designing tutoring interfaces. In this system, the tutor can—in fact must—interrupt the student as soon as he or she has produced an error (e.g., typed in an incorrect answer to a question).

In this system, then, at least one of the parties has the ability to initiate repair close to the source of the trouble.[4] The problem with this system is how it has defined *close:* It positions the correction immediately following the errorful word. Although this approach represents an interesting attempt at modeling repair, it does not incorporate what is now known about repair in tutoring.[5]

Other Correction. In chapter 5 we saw that tutors and students manage to negotiate correction in such a way that students are given the opportunity to (a) correct their own mistakes, and failing that, (b) at least initiate correction that the tutor can then accomplish, and failing that, (c) the tutor can initiate (and potentially also accomplish) correction. As in everyday conversation, the tutor withholds correcting a student until the student has passed opportunities to initiate correction. In other words, tutors do not behave the way Anderson et al's current system would expect: They do not immediately correct an error as soon as it is produced.[6] They wait until slightly after the student's errorful turn appears to be complete, and then, if the student still shows no signs of correcting his or her error, they suggest that some correction may be in order.

OTHER EFFECTS OF BANDWIDTH

In comparing our face-to-face sessions with the terminal-to-terminal sessions, our findings were generally in keeping with the literature on the differences between spoken and written English (see, e.g., Akinasso, 1982; Biber, 1986; Fox, 1987b; Rubin, 1980). We found, for instance, that in the narrow bandwidth condition (i.e., terminal-to-terminal) the tutor and student used more explicit linguistic specifications than in the broader bandwidth condition. They were less likely to use bare deictics like *this* and *that* in the narrow bandwidth condition, more likely to use complete sentences, and were less likely to use semantically empty placeholder words like *thing*. Sentences tended to be more syntactically integrated (cf. Chafe, 1982) in the former condition than in the latter, and words

---

[4]Of course, this system does not allow for the possibility that the student might need to repair something generated by the tutor.

[5]Anderson et al. (1987) did propose as an alternative method of waiting until the student has finished a complete answer before starting correction. This alternative is fully in keeping with the suggestions below.

[6]In fact, they rarely overtly correct a student's error at all.

from a more formal register (Halliday, 1973) tended to occur more in the former. Compare the following utterances, the first of which is from the broad bandwidth condition, and the second from the narrow bandwidth:

(10) T: Oh, excuse me, excuse me, I was looking at this one, and you're still working on this.

(11) T: You are perfectly right. I described the position of A wrong.

Notice in particular the affect and repetition in (10), along with the contraction of *you're*. In (11) we see words from a more formal register (*perfectly*), lack of contraction of *you are*, and the explicit formulation of what the tutor had described incorrectly.

Anaphoric devices (e.g., pronouns or full noun phrases) for tracking referents showed significant differences across the two modalities. In the face-to-face sessions, for example, a referent mentioned by one participant could be referred to with a pronoun by the other participant:

(12)
T: Like you don't *really* need to know the angle.
S: I didn't need to know it was (0.5) twenty seven?

The student's *it* refers back to the tutor's *the angle*—what we could call *cross-turn pronominalization,* inasmuch as the pronominalization crosses conversational turns. If we include demonstrative pronouns (e.g., *this*) here as pronouns, we find that cross-turn pronominalization is extremely prevalent in the face-to-face sessions, but essentially non-existent in the terminal-to-terminal sessions: I found one instance of cross-turn pronominalization in the terminal-to-terminal sessions, given as (13):

(13)
S: How do I find time it takes light to pass through the box?
T: You can figure this out.

In all other instances of cross-turn references in the terminal-to-terminal sessions, the tutors and students used full noun phrases to accomplish the reference, as in (14):

(14)
S: Here's one I lost a point on due to a sign error, copy it down on paper so you can work it while I type in my work.
T: ok show me the problem and I will write it down.

With the exception of the example given in (13), then, third person pronouns (including demonstrative pronouns) were restricted to same-turn referents, usually in a very simple syntactic configuration, as in (15):[7]

(15)
T: If a carbon has four different groups connected to it, how many stereoisomers are there of that molecule?

As we saw in chapter 6, the timing of the student's participation is used by the tutor in face-to-face dialogue to diagnose the student's level of understanding of the ongoing discussion. Given that such sensitive timing is no longer available to the tutor in the terminal-to-terminal condition, students tend to be more explicit in discussing their own assessment of their understanding in this situation. Consider the following two passages. The first is from a face-to-face interaction; the second from a terminal-to-terminal session:

(16)
T: (reading from the textbook) How close must two electrons be if the electric force between them is equal to the weight, (0.7) of either at the earth's surface.
    (0.3)
T: Okay, so what did you:

(17)
S: problem two: Person w/ eyes 1.48m above ground stands 2.4m in front of mirror whose base is .04m above ground. Find horiz. distance to base of the wall supporting the mirror of the nearest point on the floor that can be seen reflected in the mirror.
    no clue. i know the angles of incidence and refraction are the same and i can use the pathegorean theorem somehow but dont know how to start.[8]

In passage (16), the tutor watches the student carefully in the (0.3) silence and sees that the student is not making a move to begin working on the problem (or talk about how she had tried to solve the problem). Although the student says nothing in that time, the tutor makes a judgment about the student's ability to continue alone on the problem. In passage (17), on the other hand, rather than communicating her confusion with silence, the student says that she does not

---

[7]There are also nonreferring pronouns in other kinds of constructions, most of which are "dummy" uses of *it* (or other idiomatic expressions, such as *Got it!*).

[8]We have tried as much as possible to reproduce exactly what was written by each participant. Some cleaning up of typos and run-together words has been done.

know how to begin working on the problem. Such overt statements did not occur often in the face-to-face dialogues but were fairly common in the narrow bandwidth condition.

An obvious difference in the two bandwidth conditions is the opportunity for expression of affect. In the broader bandwidth condition, tutor and student can express confusion, surprise, excitement, disappointment, and so on through a variety of channels. In addition to the purely linguistic channel, they have at their disposal voice quality and intonation, facial expression, body movements, and nonlinguistic verbalizations. In the narrow bandwidth condition, tutor and student are limited to elements that can be typed from a standard keyboard.

We expected, given this limitation on communication, that tutors and students would simply abandon any attempt to convey affect. We found, on the contrary, that they invented ways within the limited means provided them to express a surprisingly wide range of emotions. They showed irony ("haha"), they laughed ("I have to laugh at these equations"), they hesitated (". . ."), all with normal alphabetic conventions. They used exclamation marks to show amazement ("I have annoying red pen marks declaring 'math error!'") and excitement ("by George, I think he's got it!"). Our speculation about this phenomenon, supported by studies in the learning literature, is that affect is important to establishing a positive learning environment for humans. Thus, even when the conditions are not conducive to displays of emotion, people work to overcome those limitations.

In large part due to the turn-taking system we imposed on the tutors and students in the narrow bandwidth condition, the kinds of collaboration they engaged in were different than in the face-to-face situation. In the broader bandwidth condition, tutors often completed and adjusted confused student statements, without the student having to ask for such help, as in the following passage:

(18) S: So I tried to look at the w*ei*ght,
     (0.9)
   T: And all's they give you is the m*a*ss.
   S: And it- yeah (0.7) Oh, that's what it was, it was the m*a*ss.

The comma at the end of the student's first line indicates falling but nonterminal intonation, suggesting that she is not finished speaking. But the fairly long silence that follows this utterance suggests to the tutor that the student is having difficulty formulating the next step; hence, without being overtly asked, the tutor completes the sentence for the student. Tutor and student also draw diagrams together, construct formulae together, figure out how to start on a problem together, and so on.

In the narrow bandwidth condition, at least as we set it up in our project, it was virtually impossible for the tutor to complete a student utterance, because no interruption of any kind was possible. Moreover, because they were not physically together, and because it was not possible to maintain an image on both terminal screens at the same time (another failure in the design of our system), there was little opportunity to create shared objects (diagrams or formulae). Creation of shared objects was possible only if one of the parties bothered to send a complex non-linguistic object at all (not an easy task, given the fact that our terminals were not equipped with graphic capabilities), and if both parties copied down the images from the screen onto a piece of paper, a strategy in fact adopted by several of the subjects in our study; or, if one gave explicit instructions to the other about how to construct a diagram or formula:

(19) T: The spot closest to the mirror that you can see lets call A. Draw a line from A to the persons eyes. This defines the angles of incidence and reflection.

The tutors and students in this study worked to overcome this limitation in other ways as well. They set up explicit common definitions of critical concepts, and, as noted earlier, students in particular were more likely to give explicit statements of their ability to proceed with a given problem:

(20) S: I was thinking that to see A through the mirror you would look at the bottom of the mirror .04 above the ground. then you have two sides of a triangle, the distance to the mirror and the vertical distance= 1.48m to eyes- .04m above the ground
   T: You are perfectly right. I described the position of A wrong. Pt A is on the mirror on the bottom edge. Now draw a line to the eyes. This defines the angle of incidence.

Again, because of the design of the system in the narrow bandwidth condition, timing was not available as a source of diagnostic information for the tutor (or the student); and, as a result of this fact, silence (i.e., time during which one of the parties sat staring at a blank screen waiting for a message from the other party) was multiply ambiguous. Some effort was thus expended by the tutors and students to make explicit the meaning of the silences:

(21)
T: You are only concerned with focusing the light. I am thinking.
S: Me too.

Here both tutor and student indicate that the silence between turns is not because they are confused, didn't receive the last message, or that the program has crashed (as it did a few times), or they are engrossed in another problem, or have gone to the bathroom; some knowledge of what the other person is doing when they are not visible is crucial for establishing an accurate model of that person. So here again we find that the tutor and student accommodated to the limitations of the system imposed on them in ways important to the design of cooperative systems.

CONCLUSIONS

In looking at the differences between face-to-face and terminal-to-terminal sessions, we see again the extreme context-sensitivity of human linguistic behavior, and the ability of people to adapt to a variety of communicative arrangements, even those not particularly conducive to effective communication. This chapter has explored some of the strategies that tutors and students use to accommodate a particular terminal-to-terminal environment; further research needs to be conducted on other terminal-to-terminal environments to determine if these strategies occur or if they are specific to one environment.

9
Indeterminacy and Rules

INDETERMINACY

The principle of indeterminacy has been a theme throughout this study. Because it is a difficult and highly theoretical notion, this chapter provides a more detailed discussion of indeterminacy, and the implications of indeterminacy for notions of rule-governed behavior. This chapter is thus an elaboration of concepts already touched upon in earlier chapters; readers who feel so inclined can skip this chapter and go directly to chapter 10.

In chapter 1, I sketched out a claim that all human conversational utterances are inherently indeterminate, in the sense that the linguistic utterance itself does not uniquely specify a single interpretation; rather, all utterances are open to multiple interpretations.

One source of this indeterminacy is the intrinsic underspecifiedness of lexical items:

> Consider our example, "That's a nice one," being said by a visitor to a host while both are looking at a photograph album. The referent of "that's" is established by the visitor pointing to a particular photograph, but what is the meaning of "nice" here? It is obvious that "nice" could mean a number of things in such a context. The visitor could be admiring the composition of the shot, or suggesting that the photograph was a good likeness of the host, or indeed that the host looked particularly well in the photograph. Whichever sense of "nice" was intended is certainly not available from the utterance alone, but remains to be made out by the host in the light of the specifics of the photograph. Here, then, the "corresponding contents" invoked by the term *remain to be discovered* by an active search of the referent. Quite a differently organized search will be initiated if the utterance is directed by a

girl to her boyfriend in front of a jeweller's window. . . . And different properties again will be looked for in a greengrocer's lettuce described as "nice." (Heritage, 1984, p. 143)

Heritage also said, "this suggests that the boundaries of the applicability of a term will be indeterminate, negotiable and subject to change" (p. 145).

Another source of indeterminacy is syntactic constructions. For example, the various *get*-constructions in English (like *get dressed, get fired, get oneself pregnant*) indicate that the subject of the clause is affected by the action, but leaves unspecified to what extent the subject is also agentive (see Arce-Arenales, Axelrod, and Fox, in press). In *Jose got dressed,* for example, the subject (Jose) is fully agentive—he is the one carrying out the action of dressing, as well as being its recipient. But in *Jose got fired,* Jose is not directly responsible for the firing (he did not fire himself), although he may have done something to bring it about (e.g., been too left-wing); in *Sally got herself pregnant,* Sally is somewhat agentive, but we also know from the "real world" that Sally did not get pregnant all by herself. The level of agentivity of the subject can only be worked out in context, and then, just as with lexical items, it is open to negotiation and reinterpretation. Even though an utterance occurs in a particular context, its meaning is not fixed, although it is situated.

Another syntactic source of indeterminacy is a phenomenon I call *syntactic blending,* whereby a speaker begins with a particular syntactic construction and then pivots on one of the words in the construction into another syntactic construction; the resulting utterance is impossible to parse into a single syntactic representation and is therefore indeterminate in its syntactic representation and its corresponding semantic representation. An example of syntactic blending is given here:

(1)
T: and you look over (at) your equation and you go okay well that's (0.2)
⇒ work and energy are the same thing.
S: Aha

In (1), the tutor seems to begin a simple copular construction (i.e., with the verb *be*) "that's work" but then pivots on the word *work* into "work and energy are the same thing." Notice that the resulting utterance, "that's work and energy are the same thing," is not a single syntactic construction (and is thus not grammatical), but it is nonetheless apparently interpretable by the student, by some mechanism we do not yet fully understand.

A further syntactic source of indeterminacy is the common practice in conversation of producing incomplete syntactic units. Such units may afford a normal syntactic parsing, by virtue of the fact that the hearer can project what the speaker was going to say (i.e., what kind of unit was being produced and how it was going to be completed). But the "void" at the end of the utterance obviously provides minimal constraints on its own interpretation, although the hearer may be perfectly capable of inferring exactly what would have been said in that place, and is thus maximally indeterminate as to its interpretation:

(2)
T: How close must two electrons be if the electric force between them is equal to the weight, (0.7) of either at the earth's surface.
  (0.3)
⇒ T: Okay, so what did you
  (0.5)
S: So this is what- we're allowed to have our little um, (0.4) sheet of formulas

The tutor's utterance "so what did you" is incomplete and hence indeterminate with regard to how it "would have" been completed. Did the tutor mean, "So what did you do?" "So what did you think about this?" or "So what did you do to start solving this problem?" Because syntactic constructions in principle allow infinite recursion, the possible completions are limitless.

Like the void in incomplete utterances, silence is obviously indeterminate in its interpretation, although not in the least impenetrable to interpretation:

(3)
S: So I tried to look at the weight,
⇒   (0.9)
T: And all's they give you is the mass.

The nine-tenth of a second of silence in this fragment has no fixed interpretation, even in the exact context of utterance. Does it "belong" to the student while she is trying to complete her utterance? Does it "belong" to the tutor while the student waits for the tutor to offer help? Or does it first belong to the student and then get re-interpreted over the course of the fragment to belong to the tutor (for related phenomena see Fox, 1987a; Heritage, 1984)? And even if we could assign ownership to the silence definitively, what would be its significance? These are matters of interpretation for the tutor and student, interpretation that is contingent and susceptible to change.

Although most linguists and philosophers have viewed the possibility of indeterminacy with distaste, regarding such a system as fuzzy and intractable to formal mechanisms, phenomenologists (Garfinkel, 1967; Wittgenstein, 1958) have embraced the functionality of indeterminacy:

> Many discussions of language use which have been stimulated by Garfinkel's observations take over the logico-linguistic assumption that the indexical features

of natural languages are an inherent defect.... Viewed from the standpoint of the present discussion, such a view is deeply mistaken. For it is by virtue of our contextualizing activities as hearers that, as speakers, we can make good sense with a reasonably small vocabulary whose terms embody a network of criss-crossing resemblances.... The indexical character of natural language use is thus a positive resource which we exploit, rather than a defect which, ideally, should be eliminated. (Heritage, 1984, pp. 149–150)

Just as this indeterminacy is a "positive resource" in everyday language use, it is exploited in tutoring as a device for achieving higher levels of student performance. The most obvious case of carefully exploited indeterminacy in our tutoring sessions is the locally managed sense of the "plans" negotiated at the openings of the sessions (see chapter 4).

For example, in the following passage, the tutor and student work out an approach to the session whereby they will "pick problems and work them." The student then contextualizes the "problems" to be picked from by bringing in to the setting a group of problems that she has tried to solve unsuccessfully on her own:

(4)
T: Then the *o*ther thing is *I* usually find for studying for tests the b*e*st thing to do is just to- just pick problems, //and work them.
S: Yeah, that's what I've been d//oing. (More or less)
T: And then *a*ny concept in the problem we can like go off and talk about it for a b//it, then go back and work some.
S: Aha
S: Okay, well we could do it from that angle then
S: And I've been doing *e*xtra problems. And I brought the ones that I've been having a hard time figur//ing out.
T: Great.

The sense of the plan at this point is thus to work problems that the student has been having "a hard time figuring out." Later in the session, the tutor discovers that the student has a copy of a sample exam for her upcoming exam. The gloss "picking problems and working them" now is used to encompass a slightly new activity—working problems the student has not seen before, but that are similar to problems she will be called on to solve on her exam. Even if the tutor and student had contextualized the phrase "picking problems and working them" in a fairly particular way early on in the session, the phrase can now be reconstructed to have included this new activity, "all along and in the first place" (Garfinkel, 1967). It is important to remember, then, that even situated utterances, which are given some interpretation at the time of their utterance, are not fixed in their meaning.

Heritage's "that's nice" example can be used to illustrate this point further. Let us say that the speaker meant that the person in the photograph looks good, and that is the meaning that the hearer "assigns" to the utterance. But is it the person's clothing that looks good? Her hair-do? Has she gained weight? Is she radiant with love? Now what if the speaker meant to comment on the person's clothing. Is it the fit of the pants that looks good? The blouse? The black suede pumps? Now what if the speaker meant that it is the blouse that looks good. Is it the color that is so attractive, setting off her skin color? Is it the cut of the blouse? The mother-of-pearl buttons? And what if the speaker meant it is the cut of the blouse that looks good. Is it the cut of the sleeve? The neckline? The placement of the darts? and what if the speaker meant the neckline? . . .

It is easy to see from this example, exaggerated as it is (see Garfinkel, 1967, for a discussion of his now-famous breach experiments, which involved a similar task), that although it may appear that an utterance in context is given a particular interpretation, that interpretation is open to an essentially indefinite amount of telescoping and zooming, and it is this open-texture that gives conversational interactions their characteristic fluid character:

> Thus at any given point both the "sense" of the utterances and "what was talked about" with the utterances remained "specifically vague" and open-ended with respect to "internal relationships, relationships to other events and relationships to retrospective and prospective possibilities" (ibid.). (Heritage, 1984, p. 95; citing Garfinkel, 1967)

Given that any utterance is indefinitely elaborable in just the way "that's nice" is, how is it that we feel we have understood an utterance at all? It is because, as Schutz (1962) suggested, we come to an interpretation that is "sufficient for all practical purposes" (p. 12), and we expect our interlocutors to be able to telescope out or zoom in from that vague but sufficient interpretation as the context requires. Thus, if for good pedagogical reasons the tutor wants the student in the previous passage to go over the problems in her sample exam, the tutor can assume that the student will retroactively hear "pick problems and work them" as including the activity they are about to engage in, as part of this process of reshaping indeterminate meaning.

As we saw in chapter 6, tutors often use the student's ability to complete a tutor-sentence as part of the diagnosis process. In a similar vein, I found that tutors often exploited the indeterminacy of (even situated) language to find out how well the students were understanding the task at hand. It seemed in many cases, for example, that the tutor waited to see what sense the student would make of a particular word or sentence that the tutor had produced, and use that information to determine how much, and in what way, the student was understanding:

(5)
T: So the main thing here, I mean, when you look at that, what is electron volts, what kind of a

```
    S:  It'//s ahm,
⇒ T:  what are we talking about.
        (1.5)
    S:  Isn't it the charge of an electron times?
        (0.9)
    T:  Right, // but what is it- what is that. Is it
    S:  The voltage?
    S:  It's smaller
    T:  No- okay, I'm no- I'm a I'm a I'm not asking a specific enough question.
```

The tutor first produces an utterance ("what is electron volts, what are we talking about)" that is fully indeterminate, for there are an infinite number of ways to describe an object or person (my colleague, a friend, the tall one over there, my old running partner, Bill's wife, a biblical scholar, a resident of North Carolina, etc.; see Schegloff, 1972).

In the same way, asking for a description of electron volts is a nondelimited task whose answer requires a display of understanding of what the tutor is after. In the previous passage, the tutor asks "What is electron volts . . . what are we talking about." These sentences are indeterminate, in that they allow for a potentially infinite range of specific descriptions of electron volts. The student starts to give one specific description: an equation for electron volts ("isn't it the charge of an electron times . . ."); this answer suggests that the student thinks the tutor is asking about setting up equations, as a starting point for working the rest of the problem. But the tutor turns out to be interested in another description: electron volts are units of energy. The student has thus given a specific interpretation to the tutor's indeterminate utterance in a way that reflects something about how the student conceives of the problem-solving process. The tutor can thus use what the student has made of the tutor's indeterminate utterance to see how the student approaches this kind of problem, and thus what her level of understanding is. In other words, the tutor can make diagnostic decisions on the basis of how the student understands the tutor's utterances, an understanding which reflects the student's ability to contextualize an indeterminate utterance *the way the tutor would* (a kind of model-matching approach). In fact in some instances it seemed that the tutor was being vague to an extreme, to see exactly what the student would do, left with that level of contextualizing to do.

Indeterminacy in the students utterances is also exploited to full tutoring advantage. Given the fact that students often formulate vague and somewhat incoherent questions, successful tutoring becomes contingent on the tutor's ability to *find a sense* for those questions such that they can be answered in a helpful way. The student can then watch the tutor's contextualizing work, by which the tutor transforms the student's somewhat incoherent question into a question that sounds like "talking physics" (or chemistry, or math, etc.). Through this process the student can learn more about the relationships between his or her own understanding of the domain and the tutor's understanding:

(6)
S: That's- that's what I don't understand. I thought (0.7) in reading the chapter and listening to his lecture and stuff, (1.0) that (0.8) potential energy, could only be measured as a- a change.
   (0.8)
T: Ah//a
S: or something. And then I asked my TA, briefly during our res- recitation this morning and he said oh, well they take (0.8) potential energy to be zero at infinity
T: Right
S: so you use that. And I don't know really how to (0.2) plug that in.

Although the student's formulation of her question is vague, eventually the tutor is able to explain potential energy to the student and to clarify the TA's apparently cryptic remark about potential energy, in the process modeling for the student how to discuss physics concepts in the language of a physicist.

So far we have seen that indeterminacy is a crucial design feature of language, and human communication, and that it is functional, and that it is exploited by both tutor and student in the construction of successful tutoring interaction. In the next section, I explore the implications of this omnipresent indeterminacy for the status of the analytic object "rules."

RULES

Throughout this study I have described recurrent patterns of tutoring interaction. But nowhere in this discussion have I posited tutoring rules that would both describe the participants' behavior and presumably stand as a representation of the internalized rules that govern their behavior. I raise this point here because several studies of tutoring within the AI community (including Clancey, 1987; Galdes, 1990) have described their findings using discourse rules.

There are several reasons for the lack of explicit rules here, and these are important enough to merit some discussion. Indeterminacy is central among them.

Discourse rules for tutoring have, in the human–computer interaction literature, often taken the form of production rules, with the *if* segment (the protasis) specifying input to be detected and the *then* segment (the apodosis) specifying the output. An example of such a discourse rule for tutoring is given below (taken from Galdes, 1990, p. 459; S = Student):

(7)

If future evidence → S did not forget something *then* drop this topic (because it is not an error).

There are several things to notice about rule formulations of this sort. First is the relationship between this rule and actual situated tutor behavior: in what relationship does this rule stand to any given tutor's behavior at a given moment? Is it meant to stand for some internalized representation that determines tutor behavior? Or which will be used to determine a computer tutor's behavior? Second, and closely related to the first, is the indeterminacy of the terms of the rule: How will the follower of the rule know what *future evidence* is, or what it means to *drop a topic,* or how to know when future evidence suggests that the S[tudent] *did not forget something* (and what is *something* in this case?)? Third is the question of context-sensitivity: Does this rule always apply when the protasis is satisfied, no matter what the surrounding discourse context (assuming for the moment that one can determine outside of a particular context when the protasis is satisfied)? I address each of these questions here.

The relationship between a rule and the behavior it is meant to represent has been a topic of study since the first studies of human behavior. The traditional approach sees a rule as determinative of behavior; that is, people *follow* rules:

> In essence, the "rule-governed" model of human conduct is a very simple one. It begins from the presumption that human actors are generally equipped with an array of rules which they "follow" (or by which they are "guided" or "governed") in situations of action . . . . the traditional model of "rule-governed" conduct works in the following fashion. The actors are treated as encountering a situation of action to which one or more of the rules they have learned or internalized "apply." Their actions in this context are then analysed as "guided" or "caused" by the rules which they have previously acquired. (Heritage, 1984, pp. 104–105)

The production rule model falls within this description of the traditional model of "rule-governed" behavior, by virtue of the following assumption: That the actor involved (either a human tutor or a computer tutor) looks for the situation described in the protasis, and, if that situation holds, carries out the action described in the apodosis.

In contrast, an ethnomethodological treatment of rules and behavior (see, e.g., Garfinkel, 1967; Heritage, 1984; Suchman, 1987) sees apparently rule-governed behavior as arising through norms toward which people orient in interpreting and creating the contexts in and through which they act. The normative, rather than determinative, nature of behavior can be seen in the consequences of violations: If it were the case that rules determine behavior, then violations should be uninterpretable, just as in traditional views of syntax a string of symbols that do not follow the rules of syntax for that language is ungrammatical (or uninterpretable to the grammar).

In fact, violations of "rules" (or norms) offer rich material for interpretation. For example, suppose there is a rule (or norm) that students initiate discussions of their troubles at the openings of tutoring sessions, and suppose that one of the students in our study failed to do that; it is unlikely that the tutor in that session would find the student's behavior simply "ungrammatical." Rather, the tutor would use the norm to interpret what would then become, by virtue of that norm, the "student's silence," and would attribute to the student one or more motivations for the silence—the student is shy, doesn't know how to describe her or his trouble, is out of breath from running, is uncomfortable with a female tutor, and so on. The tutor would then use one or more of these interpretations as the basis for her next action, which would almost undoubtedly be some kind of prompt to the student to get the student to talk about what brought the student for tutoring, or how they should proceed with the tutoring (see chapter 4 for confirming evidence for this claim).

Now notice that we could try to formulate this turn of events with a production rule, of the following sort:

(8)
if the student initiates trouble-talk,
   then respond to it

ELSE
if the student remains silent,
   then initiate trouble-talk

According to this rule, a tutor encountering a student who remains silent at the opening of a session would initiate discussion of what brought the student for tutoring.

But, and this "but" brings us to our second point, how does the tutor know when the student is being silent, rather than, say, swallowing the last bite of a sandwich, looking for a page in the textbook to start with, or tying her or his shoe? Let's say we try to formulate our rule more specifically, to include an operational definition of silence:

(8')
if the student is not engaged in any non-verbal activity,
       and
if no linguistic utterances are being made by the student,
   then initiate trouble talk

So if the student is just finishing lunch, or looking out for a bee that has gotten into the room, the rule will presumably not apply, because these are nonverbal activities. But how does the tutor know exactly what counts as *nonverbal activity*? Does the student nervously tapping his or her foot count as nonverbal activity in this case (i.e., should the tutor wait for it to end to prompt the student to talk)? What if the student looks as if he or she is "thinking hard" and is working on trying to figure out what to say—is that activity at all?

The answer to these questions is that we have found ourselves in the infinite underspecifiedness of rules, to wit: Because each term in the rule is indeterminate (i.e., there is no way to pinpoint for all possible contingencies exactly what will count as covered by that term and what will count as outside that term) then the rule cannot determine behavior, because the rule can never say for itself "you must apply me here" (it can never, as Garfinkel said, "step forward and claim its own instances"), inasmuch as deciding whether "here" satisfies the protasis can never be fully specified in the rule (see Suchman, 1987, for a similar discussion). So the pervasive indeterminacy of language and communication eliminates the possibility that human behavior is rule-governed (in the sense of determined by rules).

This point about rules and their relationship to the behavior they describe is closely related to what Garfinkel has called "monsters":

> You can start with a command . . . "Take a game, any game, write down the instructions as to how to play that game. Finished?" Then you pass it to somebody else. That other person is asked: "Do you have the instructions to the game? Now find monsters in those instructions so that if you needed to be instructed in that way you couldn't possibly make it out." Say we are going to propose a game of ticktacktoe. Two persons play ticktacktoe. Any two persons? When, today? Tomorrow? Do we have to be in sight of each other? Can we play by mail? Can one player be dead? . . . My classes can tell you that creating such problems is the easiest thing in the world. It comes off every time without fail. (Garfinkel, 1968, cited in Heritage, 1984, pp. 124–125)

As Garfinkel made clear in this passage, there are an infinite number of such "monsters" lurking behind every rule formulation:

> In sum, no matter what the circumstances, the clarity of their delineation and the transparency with which their particulars fall under a rule, there always remains an open set of unstated conditions of the rule's application. (Heritage, 1984, p. 126)

Now we could try to get out of this action-description problem by formulating the production rule in terms of goals, rather than in terms of behavior:

(8″)

if the student does not have the goal of initiating trouble-talk,

    then initiate trouble-talk

But this formulation pushes the difficulties sketched here into another corner, as it begs the question of how the tutor recognizes when the student has the relevant goal—presumably the tutor would recognize the student's goals on the basis of the student's behavior, bringing us back full circle to the intransigent difficulties with rules as a model of tutoring behavior (see Suchman, 1987, for a cogent discussion of the problems with goal recognition in human–computer communication).

We now come to the third reason for being cautious with rulelike formulations of discourse patterns. It is unclear if such rules are meant to apply in all contexts or whether they are only meant to apply under certain specific conditions. Presumably, if no condition is mentioned in the rule, then the rule-follower must apply the rule in all contexts for which the protasis is satisfied. But, as I previously mentioned, rules (or norms) are often either violated, for some local reason, or are not even invoked in certain settings. For example, to borrow an illustration from Heritage (1984), the rule (or norm) to greet an acquaintance can be violated (one can fail to greet a friend, although the friend is thereby licensed to infer all manner of evil intentions), and it can fail to be invoked (one doesn't greet a roommate who has returned to the livingroom from getting a cupcake in the kitchen, or if one does, it licenses "special" inferences, like that one wants a bite of the cupcake). Similarly, although there may be a norm that says that students should open a tutoring session with a discussion of what inspired them to seek tutoring, such a norm is clearly violated in one of the sessions in which the tutor prompts the student ("where shall we start"); and the norm is not even invoked when the student and tutor meet for a second time, as was the case in the present study.

This chapter has elaborated the fundamental notions of indeterminacy and nonrule-governedness. Readers interested in additional discussion of these issues should consult Garfinkel (1967), Heritage (1984), and Suchman (1987).

10 Conclusions

CHARACTERISTICS OF HUMAN–HUMAN TUTORING

In this study, we have explored the properties of human–human tutoring. This chapter summarizes our basic findings on the characteristics of human tutoring.

The first finding, discussed in detail in chapter 4 and threaded throughout the other chapters, is that the function of tutoring is to teach the student methods for contextualizing abstract symbols and descriptions—including linguistic phrases in problem statements, formulae, constants, variables, and so forth. Tutoring is thus seen not as a transfer of information about a scientific domain, but as a way of teaching methods and practices for using static problem statements as a bridge to a series of activities.

The importance of this property rests on two facts: First, the contextualization is individually tailored, to exactly fit the needs of the student at the moment; and second, contextualization makes visible to the student the otherwise hidden activities that are foundational to scientific (and mathematical) concepts. The value of this latter fact from the perspective of situated action (Suchman, 1987) has been discussed; there is further cognitive value for this fact, as can be seen in recent work on the subject of embodied knowledge and metaphor (e.g., Lakoff, 1987; Lakoff & Johnson, 1980); in this work, the vital link between abstract concepts and descriptions and the bodied activities which lie behind those forms is made explicit. For example, H. Collins (1988) described the vast discrepancy between the methods section of a scientific paper (which is intended to enable any other scientist to replicate the procedures described therein) and the real-world physical actions that a scientist must perform in order to carry out the procedures so described; the linguistic description is a kind of conventionalized shorthand for those specific actions, and someone learning to be a scientist must be taught how to relate the shorthand to real-life activities in setting (to use Lave's term; see Lave, 1988). Moreover, as the research on metaphor suggests, our ability to understand complex abstract concepts is directly related to our ability to think of those concepts in concrete, often body-based, terms (thinking of time in terms of physical space, as experienced by the asymmetrical orientation of the human body, for example). It thus appears that at its core our knowledge—of even extremely abstract systems—is organized in terms of real-world physical actions on concrete objects; in making explicit the actions-on-objects that are hidden by scientific problem statements, then, tutoring builds a crucial educational bridge for students.

The second finding is the indeterminacy of tutoring language and tutoring communication. That is, in tutoring (as in all conversation) a given linguistic item (including silence) is in principle open to an indefinite number of interpretations and reinterpretations. Indeterminacy is a fundamental principle of interaction and can be seen as at least a partial cause of the nonrule-governed nature of interaction (see especially chapter 9).

The last finding regarding the nature of tutoring is that it is entirely interactive, involving complete mutual orientation: Every utterance, every explanation, every question of clarification, is designed for a particular student or tutor, with particular needs, on a particular occasion, with a particular discourse history, as displayed and understood by the tutor and student. Every utterance, every problem-solving sequence, every session, is a joint achievement by both people.

CONCLUSIONS

The goal of this study was to explore the systematic methods and practices by which tutoring is accomplished. It has not been the intention of this work to speculate on the particular uses in particular systems to which the findings presented here can be put.

Nonetheless, while remaining neutral on this issue, I would like to raise now some of the considerations that could be brought to bear in understanding the implications of these findings for system design.

As we saw in the preface, it is of serious theoretical and practical concern in human–computer interaction to determine what can be simulated from human communication (if anything) and to determine what can make human–computer interaction (as opposed to human–human interaction) "flow more smoothly" (Galdes, 1990, p. 373). These determinations cannot be made solely by examining human–human interaction, because human–human interaction by itself does not expand our knowledge of the asymmetries between humans and computers. The findings presented here, then, need to be taken as one side of an equation; the other side consists of theoretical and practical explorations of computers and their interactional properties (e.g., Suchman, 1987). In utilizing the findings presented here in the design of interactive systems, the following issues may need to be addressed.

First, the theoretical limitations of systems: What are the constraints imposed by the symbol-processing architectures of interactive systems? As Suchman (1987) and Dreyfus and Dreyfus (1986) made clear, the constraints imposed by a lack of situated interpretation and a lack of human body experiences make simulation of human activity highly problematic. We have seen in this study that linguistic utterances are indeterminate and require contingent interpretation based on the context (recall, e.g., the discussions of the interpretation of silence); is such interpretive ability achievable in intelligent systems?

Second, once we know the limitations of intelligent systems, we need to determine the strengths of computers rather than assume that we should faithfully simulate human behavior: How can we make the best use of intelligent systems given their particular strengths and limitations? This study has presented findings on the behavior of human tutoring; which of these findings are appropriate to implement in a system, and which are feasible but not appropriate? This is also a theoretical consideration, inasmuch as it requires a careful analysis of the asymmetries between computers and humans and an assessment of the relative strengths of intelligent systems.

Third, there is the issue of feasibility of implementation at the current time: Is the implementation, for example of a flexible turn-taking system, possible given current technology, and are the benefits of such a change worth the cost of the implementation? This is a purely practical consideration that can be determined both on a case-by-case basis for each system and on the level of the state-of-the-art for the field, after the first two issues have been dealt with.

Fourth, if simulating a particular human activity is appropriate but not feasible at this time, what alternative resources can be brought to bear to accomplish roughly the same task? For example, if it is appropriate but not feasible to model human processes of repair, what can be modeled to capture at least some of the functionality of repair? For a given system, this too is a practical consideration that can be determined on the basis of the idiosyncracies of that system; at the level of the field, this consideration requires a systematic evaluation of the current capabilities of computer tutors against a backdrop of a systematic evaluation of what is appropriate.

REFERENCES

Akinasso, N. (1982). On the differences between spoken and written language. *Language and Speech, 25,* 97–125.

Anderson, J. R. (1988). The expert module. In M. Polson & J. Richardson (Eds.), *Foundation of intelligent tutoring systems* (pp. 21–53). Hillsdale, NJ: Lawrence Erlbaum Associates.

REFERENCES

Anderson, J., Boyle, C. F., Corbett, A., & Lewis, M. (1987). *Cognitive modelling and intelligent tutoring*. Unpublished manuscript.
Anderson, J., Boyle, C. F., Farrell R., & Reiser, B. (1984). Cognitive principles in the design of computer tutors. In *Proceedings of the sixth annual conference of the Cognitive Science Society* (pp. 2–9). Hillsdale, NJ: Lawrence Erlbaum Associates.
Arce-Arenales, M., Axelrod, M., & Fox, B. (in press). Middle diathesis in cross-linguistic perspective. In B. Fox & P. Hopper (Eds.), *Voice: Form and function*. Amsterdam: John Benjamins.
Atkinson, P., & Drew, P. (1979). *Order in court*. London: Macmillan.
Atkinson, P., & Heritage, J. (1984). *Structures of social action*. Cambridge: Cambridge University Press.
Baldwin, J., & Siklossy, L. (1977). An unobtrusive computer monitor for multi-step problem solving. *International Journal of Man-Machine Studies, 9*, 349–362.
Biber, D. (1986). Spoken and written textual dimensions in English. *Language, 62*, 384–414.
Bobrow, D., & Winograd, T. (1977). An overview of KRL. *Cognitive Science, 1*, 3–46.
Chafe, W. (1982). Integration and involvement in speaking, writing, and oral literature. In D. Tannen (Ed.), *Spoken and written language* (pp. 35–53). Norwood, NJ: Ablex.
Clancey, W. (1987). *Knowledge-based tutoring*. Cambridge, MA: MIT Press.
Coates, J., & Cameron, D. (Eds.). (1988). *Women in their speech communities*. London: Longman.
Collins, A., Brown, J. S., & Newman, S. (1986). *Cognitive apprenticeship: Teaching the craft of reading writing and mathematics* (BBN report no. 6459).
Collins, H. (1988). Paper presented at the annual meeting of the American Anthropological Society, Phoenix, AZ.
Coombs, M., & Alty, J. (1980). Face-to-face guidance of university computer users—II. *International Journal of Man-Machine Studies, 12*, 407–429.
Dellarosa, D. (1986). A computer simulation of children's arithmetic word problem solving. *Behavior Research Methods, Instruments, and Computers, 18*, 147–154.
Dellarosa, D., Kintsch, W., Reusser, K., & Weimer, R. (1988). The role of understanding in solving word problems. *Cognitive Psychology, 18*, 405–438.
Dreyfus, H., & Dreyfus, S. (1986). *Mind over machine*. New York: The Free Press.
Duranti, A. (1984). *Intentions, self, and local theories of meaning: Words and social action in a Samoan context* (Center for Human Information Processing Tech. Rep. No. 122). La Jolla: University of California, San Diego.
Edmonds, E. (1982). The man-computer interface: a note on concepts and design. *International Journal of Man-Machine Studies, 16*, 231–236.
Fitter, M. (1979). Towards more "natural" interactive systems. *International Journal of Man-Machine Studies, 11*, 339–350.
Fox, B. (1987a). Interactional reconstruction in real-time language processing. *Cognitive Science, 11*, 365–387.
Fox, B. (1987b). *Discourse structure and anaphora*. Cambridge: Cambridge University Press.
Fox, B. (1990). *Final report for the Human Tutorial Dialogue Project*. Unpublished manuscript.
Fox, B., & Karen, L. (1988). Collaborative cognition. In *Proceedings of the 12th annual meeting of the Cognitive Science Society*. Montreal: McGill University Press.
Gaines, B., & Shaw, M. (1986). Foundations of dialog engineering. *International Journal of Man-Machine Studies, 24*, 101–123.
Galdes, D. (1990). *An empirical study of human tutors: The implications for intelligent tutoring systems*. Unpublished doctoral dissertation, Ohio State University, Columbus.
Garfinkel, H. (1967). *Studies in ethnomethodology*. Englewood Cliffs, NJ: Prentice-Hall.
Giddens, A. (1979). *Central problems in social theory*. Berkeley: University of California Press.
Givon, T. (1979). *On understanding grammar*. New York: Academic Press.
Goodwin, C. (1979). The interactional construction of a sentence. In G. Psathas (Ed.), *Everyday language: Studies in ethnomethodology* (pp. 97–121). New York: Irvington.

REFERENCES

Goodwin, C. (1981). *Conversational organization.* New York: Academic Press.
Goodyear, P. (Ed.). (1991). *Teaching knowledge and intelligent tutoring.* Norwood, NJ: Ablex.
Halliday, M. (1973). *Explorations in the functions of language.* New York: Elsevier.
Harris, R. (1981). *The language myth.* New York: St. Martin's Press.
Hayes, P. (1983). Introduction. *International Journal of Man-Machine Studies, 19,* 229–230.
Hayes, P., & Reddy, D. R. (1983). Steps toward graceful interaction in spoken and written man-machine communication. *International Journal of Man-Machine Studies, 19,* 231–284.
Heritage, J. (1984). *Garfinkel and ethnomethodology.* London: Polity Press.
Jefferson, G. (1979). A technique for inviting laughter and its subsequent acceptance/declination. In G. Psathas (Ed.), *Everyday language: Studies in ethnomethodology* (pp. 79–96). New York: Irvington.
Kidd, A., & Cooper, M. (1985). Man-machine interface issues in the construction and use of an expert system. *International Journal of Man-Machine Studies, 22,* 91–102.
Labov, W., & Fanshel, D. (1977). *Therapeutic discourse.* New York: Academic Press.
Lakoff, G. (1987). *Women, fire, and dangerous things.* Chicago: University of Chicago Press.
Lakoff, G., & Johnson, M. (1980). *Metaphors we live by.* Chicago: University of Chicago Press.
Lave, J. (1988). *Cognition in practice.* Cambridge: Cambridge University Press.
Leontiev, A. (1981). *Psychology and the language learning process.* New York: Pergamon Press.
Levinson, S. (1983). *Pragmatics.* Cambridge: Cambridge University Press.
McTear, M. (1987). *The articulate computer.* New York: Basil Blackwell.
Miller, L., & Thomas, J. Jr. (1977). Behavioral issues in the use of interactive systems. *International Journal of Man-Machine Studies, 9,* 509–536.
Oberem, G. (1987). *ALBERT: A physics problem-solving monitor and coach.* Paper presented at Third International Conference on Artificial Intelligence and Education, Pittsburgh, PA.
Ochs, E. (1979). Transcription as theory. In E. Ochs & B. Schieffelin (Eds.), *Developmental pragmatics* (pp. 43–72). New York: Academic Press.
Ochs, E. (1988). *Culture and language development.* Cambridge: Cambridge University Press.
Osherson, D., & Lasnik, H. (1990). *An invitation to cognitive science: Language.* Cambridge, MA: MIT Press.
Pomerantz, A. (1975). *Second assessments: A study of some features of agreements/disagreements.* Unpublished doctoral dissertation, University of California, Irvine.
Pomerantz, A. (1978). Compliment responses: Notes on the cooperation of multiple constraints. In J. Schenkein (Ed.), *Studies in the organization of conversational interaction* (pp. 79–112). New York: Academic Press.
Reddy, M. (1979). The conduit metaphor. In A. Ortony (Ed.), *Metaphor and thought* (pp. 284–324). Cambridge: Cambridge University Press.
Rogoff, B. (1990). *Apprenticeship in thinking.* Oxford University Press.
Rogoff, B., & Lave, J. (1984). *Everyday cognition.* Cambridge, MA: Harvard University Press.
Sacks, H. (1992). *Lectures on conversation.* Cambridge, MA: Blackwell.
Sacks, H., Schegloff, E., & Jefferson, G. (1974). A simplest systematics for the organization of turn-taking. *Language, 50,* 696–735.
Schank, R., & Abelson, R. (1977). *Scripts, plans, goals, and understanding.* Hillsdale, NJ: Lawrence Erlbaum Associates.
Schegloff, E. (1972). Notes on a conversational practice: Formulating place. In D. Sudnow (Ed.), *Studies in social interaction* (pp. 75–119). New York: Academic Press.
Schegloff, E. (1981). Discourse as an interactional achievement. In D. Tannen (Ed.), *Text and talk* (pp. 71–93). Washington, DC: Georgetown University Press.
Schegloff, E. (1987). Between macro and micro: Contexts and other connections. In J. C. Alexander et al. (Eds.), *The micro–macro link* (pp. 207–234). New York: Columbia University Press.
Schegloff, E., Jefferson, G., & Sacks, H. (1977). The preference for self-correction in the organization of repair in conversation. *Language, 53,* 361–382.

Schegloff, E., & Sacks, H. (1973). Opening up closings. *Semiotica, 7,* 289–327.
Schenkein, J. (Ed.). (1978). *Studies in the organization of conversational interaction.* New York: Academic Press.
Schutz, A. (1962). *Collected papers.* The Hague: Martinus Mijhoff.
Searle, J. (1979). *Expression and meaning.* Cambridge: Cambridge University Press.
Sleeman, D., & Brown, J. S. (Eds.). (1982). *Intelligent tutoring systems.* New York: Academic Press.
Suchman, L. (1987). *Plans and situated actions.* Cambridge: Cambridge University Press.
Tannen, D. (1990). *You just don't understand.* New York: William Morrow.
Terasaki, A. (1976). *Pre-announcement sequences in conversation* (Social Sciences Working Paper No. 99). Irvine: University of California, Irvine.
Thorne, B., Kramerae, C., & Henley, N. (Eds.). (1983). *Language, gender and society.* Rowley, MA: Newbury House.
VanLehn, K. (1988). Student modeling. In M. Polson & J. Richardson (Eds.), *Foundation of intelligent tutoring systems* (pp. 55–78). Hillsdale, NJ: Lawrence Erlbaum Associates.
Vygotsky, L. (1978). *Mind in society.* (M. Cole, V. John-Steiner, S. Scribner, & E. Souberman, eds.). Cambridge, MA: Harvard University Press.
Wilkins, D., Clancey, W., & Buchanan, B. (1988). Using and evaluating differential modeling in intelligent tutoring and apprentice learning systems. In J. Psotka, L. D. Massey, & S. Mutter (Eds.), *Intelligent tutoring systems: Lessons learned* (pp. 257–277). Hillsdale, NJ: Lawrence Erlbaum Associates.
Wittgenstein, L. (1958). *Philosophical investigations.* New York: Macmillan.
Woolf, B. (1984). *Context-dependent planning in a machine tutor.* Unpublished doctoral dissertation, University of Massachusetts, Amherst.

Author Index

A

Abelson, R., 83, 122
Akinasso, N., 101, 120
Alty, J., 94, 121
Anderson, J. R., 51, 68, 82, 99, 101, 120, 121
Arce-Arenales, M., 108, 121
Atkinson, P., 5, 15, 121
Axelrod, M., 108, 121

B

Baldwin, J., 94, 121
Biber, D., 101, 121
Bobrow, D., 83, 121
Boyle, C. F., 51, 68, 99, 101, 121
Brown, J. S., 60, 68, 83, 121, 123
Buchanan, B., 68, 82, 123

C

Cameron, D., 10, 121
Chafe, W., 101, 121
Clancey, W., 68, 82, 113, 121, 123
Coates, J., 10, 121
Collins, A., 60, 121
Collins, H., 118, 121
Coombs, M., 94, 121
Cooper, M., 94, 122

D

Dellarosa, D., 47, 121
Drew, P., 15, 121

Dreyfus, H., 120, 121
Dreyfus, S., 120, 121
Duranti, A., 4, 121

E

Edmonds, E., 94, 121

F

Fanshel, D., 52, 122
Farrell, R., 51, 121
Fitter, M., 94, 121
Fox, B., ix, 3, 5, 12, 36, 37, 47, 101, 108, 109, 121

G

Gaines, B., 94, 121
Galdes, D., vii, 61, 113, 119, 121
Garfinkel, H., 5, 41, 50, 109, 110, 111, 114, 116, 117, 121
Giddens, A., 5, 6, 21
Givon, T., 77, 121
Goodwin, C., 20, 36, 121, 122
Goodyear, P., 51, 122

H

Halliday, M., 102, 122
Harris, R., 4, 122
Hayes, P., viii, 94, 98, 122
Henley, N., 10, 123

AUTHOR INDEX

Heritage, J., 3, 5, 6, 8, 14, 15, 20, 108, 109, 110, 111, 114, 116, 117, 121, 122

J

Jefferson, G., 5, 12, 14, 15, 16, 18, 20, 55, 77, 78, 80, 122
Johnson, M., 4, 118, 122

K

Karen, L., 47, 121
Kidd, A., 94, 122
Kintsch, W., 47, 121
Kramerae, C., 10, 123

L

Labov, W., 52, 122
Lakoff, G., 4, 118, 122
Lasnik, H., 4, 122
Lave, J., 4, 5, 6, 7, 42, 47, 85, 89, 90, 119, 122
Leontiev, A., 5, 122
Levinson, S., 15, 16, 17, 18, 20, 37, 122
Lewis, M., 68, 99, 101, 121

M, N

McTear, M., viii, 122
Miller, L., 94, 122
Newman, S., 60, 121

O

Oberem, G., 67, 68, 122
Ochs, E., 14, 47, 122
Osherson, D., 4, 122

P

Pomerantz, A., 5, 20, 37, 52, 53, 55, 58, 76, 122

R

Reddy, D. R., 94, 98, 122
Reddy, M., 4, 122
Reiser, B., 51, 21
Reusser, B., 51, 121
Rogoff, B., 47, 122

S

Sacks, H., 5, 12, 14, 15, 16, 18, 20, 55, 77, 78, 98, 122, 123
Schank, R., 83, 122
Schegloff, E., 5, 12, 14, 15, 16, 18, 20, 25, 32, 36, 55, 71, 77, 78, 97, 98, 112, 122, 123
Schenkein, J., 5, 20, 123
Schutz, A., 111, 123
Searle, J., 27, 123
Shaw, M., 94, 121
Siklossy, L., 94, 121
Sleeman, D., 68, 83, 123
Suchman, L., viii, 1, 4, 5, 36, 41, 47, 87, 94, 114, 116, 117, 118, 120, 123

T

Tannen, D., 10, 123
Terasaki, A., 20, 123
Thomas, J., Jr., 94, 122
Thorne, B., 10, 123

V

VanLehn, K., 68, 83, 123
Vygotsky, L., 5, 6, 47, 60, 123

W

Weimer, R., 47, 121
Wilkins, D., 68, 82, 123
Winograd, T., 83, 121
Wittgenstein, L., 109, 123
Woolf, B., 67, 68, 123

Subject Index

A

Activity, 85-87
  nonverbal, 115-116
  theory, 5-6
Adjacency pairs, 15, 21-23, 26
  examples of, 18
  three types of expansions of, 18-20
Adult Math Project, 7
Asymmetry
  between tutor and student, 25-30
Attribute silence, 17

B

Bandwidth
  effects of, 101-106
  turn-taking and repair, 94-101
Behavior, human
  rule-governed, 7-8, 114-115
Black box modules, 82

C

Cognition, 7
  collaborative, 47
Cognitive models, 82-83
Communication
  description of, 3-4
  face-to-face vs. human computer, 11-12, 102-104, 106
    and cross-turn pronominalization, 102
    regarding turn-taking and repair, 94-101
  in interactive systems, 4-5
  models of, 4-5
  and silence, 71-72, 74, 103-105
Competitive
  vs. noncompetitive overlapping, 16-17
Computer tutor (*see* Tutor, computer)
Conduit metaphor model, 4-5
Contextualization, 25-26, 85-89, 118-119
  beginning steps of, 43-44
  definition of, 1-2
Conversation
  organization
    of everyday, 15-20
    of tutoring dialogues, 20-25
Conversation analyses, 5-6
Correction, 27-28, 44
  distinguished from tutor assistance and repair, 52
  silence and, 54, 64
  strategies, 65-67
  and turn-taking, 94-101
Cross-turn pronominalization, 102
Cuing, 62-65

D

Declarative knowledge, 82-83
Dialogue, tutoring
  diagnosis regarding, 69-81
  interdeterminacy of, 2-3
  responsibility for, 28-30
  structure of, 15-30
Diagnoses, 80-81, 103-105

interaction and, 69-76
  metric of, 76-80
  silence and, 71-72, 103-104
  student participation and, 103
  timing and, 68-69, 72-76

E

Ethnomethodology, 5-6
Expert models, 82-83, 85-88, 91-92

G

Gap
  -closing, procedure of, 90
  silence as a, 17-18
Glass box modules, 82

H

Human behavior (*see* Behavior, human)
Human tutor (*see* Tutoring, human)

I

If-then rule system, 82-83
Indeterminacy, 119
  functionality of, 109-110
  and interpretation, 107, 111-113
  and negotiating plans, 110
  and rules, 113-117
  syntactic sources of, 107-110
  of tutoring dialogs, 203
Insert expansion, 19, 22, 24, 25
Instruction, 41
  organization of tutoring dialogues and, 23-25
Interaction
  as a diagnostic resource in tutoring, 68-81
Interactional achievements, 6
Interpretation, 7, 8, 115
  and indeterminacy, 107, 111-113
Interruption
  and repair, 99-100
Intonation, 62, 104

K

Knowledge
  declarative vs. procedural, 82-83

M

Multi-unit turns, 78-79, 96-98

N

Narrative structures
  vs. question-answer structures, 24-25
Next turn repair initiator (NTRI), 19-20

O, P

Opening sessions/segments, 31-50
Overlap
  competitive vs. noncompetitive, 16-17
Postexpansion, 19, 24-25
Preexpansion, 18
Problem solving
  diagnosis regarding, 68
  expert and student models in, 82-83, 85-88, 91-92
  negotiation of an approach to, 35-39
  organization of tutoring dialogues and, 20-23
  strategies, 83-84
Problem statements, 20-21, 88-92
  scientific and mathematical, 41-46
Procedural knowledge, 82-83
Production rules, 113-117

Q

Question-answer structure, 21
  vs. narrative structures, 24-25

opening of segments, 45-50
of student and tutor, 26

R

Repair, 19-21, 24, 27-28
  distinguished from tutor assistance and correction, 52
  and turn-taking, 94-101
Repetition, 27
Representations, knowledge, 83
Response, timing of
  and diagnosis, 68-69
  of tutor, 80-81
Rule-governed behavior, 7-8, 114-115

S

Schema systems, 83
Setting, 85-87, 92-93, 97
Social interaction, 7
Silence, 28, 64, 103-106, 115-116
  attributable, 17
  correction and, 54
  as a gap, 17-18
  display of understanding and, 71-72, 74
  of student during negotiation of a plan in opening session, 36-37
  turn-taking system and, 16-18
Simultaneous talk, 16-17
Student
  asymmetry between tutor and, 25-30
  display of understanding of, 52-55, 70-76
  modeling, 82-83
  and negotiation of a problem-solving approach, 35-39
  revealing need for tutoring, 32-35
  participation of, 76-81, 103
  tutor assistance, deciding on level of, 39-41
Syntactic blending, 108
Syntactic parsing, 108-109

T

Telementation model, 4
Terminal overlap, 16
Test-taking, 87
Timing of response
  and diagnosis, 68-69, 72-74
  of tutor, 80-81
Transition relevance place (TRP), 16, 78, 95-96
Turn constructional units (TCUs), 15-17, 98
Turn-taking systems, 11-12, 78-80
  as aspect of conversation structure, 15-18
    transition relevance place (TRP), 16, 78
    turn constructional units (TCUs), 15-17
  in classroom setting, 97
  and repair
    face-to-face and terminal-to-terminal, 94-101
  rules of, 16-18, 95-96
  silence and, 16-18, 103-106
  simultaneous talk and, 16-17
Tutor assistance
  asking questions and, 58-61
  correction and, 51-67
  deciding on level of, 39-41
  distinguished from correction and repair, 52
  face-to-face vs. separate rooms, 61-62
  intonation and cuing, 62-65
  utterance-completion strategy, 63-65
Tutor, computer
  and repair, 97-98
  versus human tutors, 11-12, 102-104, 106, 119-120
Tutoring, human
  activity and setting regarding, 85-87, 92-93
  audio- and video-taping of, 12
  bandwidth

effects of, 101-106
turn-taking and repair, 94-101
characteristics of, 118-120
correction in, 51-67, 94-101
dialogue structure in, 15-30
  asymmetry between tutor and student, 25-30
  instruction, 23-25
  problem solving, 20-23
  responsibility for, 28-30
  turn-taking system, 15-18
expert and student models in, 82-83, 85-88, 91-92
function of, 1-2
  asymmetry between student and tutor and, 25-27
gender and, 10-11
goal of, 82-93
indeterminacy of dialogues of, 2-3
interaction as a diagnostic resource in
  metric of, 76-80
  silence, 71-72, 74, 103-105
  timing, 68-69, 72-74
methodology
  data collection, 10-12
  transcription notation, 12-14
opening of segments
  question-answer structure, 45-50
  scientific and mathematical problem statements, 41-46
opening of sessions, 31-41
  negotiation of a problem-solving approach, 35-39, 110
  student revealing need for tutoring, 32-35
  tutor assistance, deciding on level of, 39-41
questions of student and tutor during, 26
rules regarding, 113-117
students displaying level of understanding during, 52-55, 70-76
theoretical background for, 3-8
  ethnomethodology vs. activity theory, 5-6
three global findings on, 1-3
timing of response
  and diagnosis, 68-69, 72-74
  of tutor, 80-81
turn-taking system in, 11-12, 78-80
  as aspect of conversation structure, 15-18
  in classroom setting, 97
  and repair, 94-101
  rules of, 16-18, 95-96
  silence and, 16-18, 103-106
  simultaneous talk and, 16-17

U

Utterance-completion strategy, 63-65

For Product Safety Concerns and Information please contact our EU representative GPSR@taylorandfrancis.com Taylor & Francis Verlag GmbH, Kaufingerstraße 24, 80331 München, Germany

Printed and bound by CPI Group (UK) Ltd, Croydon, CR0 4YY
08/06/2025
01897007-0007